The Brooks/Cole Series in Advanced Mathematics
Premier Authors Dedicated to Teaching

Series editor Paul J. Sally, Jr. and Brooks/Cole have developed this prestigious list of books for classroom use. Written for post-calculus to first year graduate courses, these books maintain the highest standards of scholarship from authors who are leaders in their mathematical fields.

"A number of years ago I was struck by the need for high quality textbooks for advanced mathematics courses. In 1997, I decided to collaborate with Brooks/Cole to create a Series in Advanced Mathematics. As editor for the series, it is my intention to develop a collection of books in the post-calculus to first year graduate study range. I will maintain the highest standards of scholarship by publishing authors who are leaders in their mathematical fields. The texts in the series should be relevant to a wide range of undergraduate and graduate students. My goal is to expand the series list with several new texts every year." – *Paul J. Sally, Jr.*

Abstract Algebra/Modern Algebra
Isaacs, *Algebra: A Graduate Course* (0-534-19002-2)
Solomon, *Abstract Algebra* (0-534-39996-7)

Advanced Calculus
Fitzpatrick, *Advanced Calculus: A Course in Mathematical Analysis* (0-534-92612-6)

Approximation Theory
Cheney/Light, *A Course in Approximation Theory* (0-534-36224-9)

Fourier Analysis
Folland, *Fourier Analysis and Its Applications* (0-534-17094-3)
Pinsky, *Introduction to Fourier Analysis and Wavelets* (0-534-37660-6)

Geometry
Isaacs, *Geometry for College Students* (0-534-35179-4)

Mathematics of Finance
Goodman/Stampfli, *The Mathematics of Finance: Modeling and Hedging* (0-534-37776-9)

Numerical Analysis
Kincaid/Cheney, *Numerical Analysis: Mathematics of Scientific Computing*, Third Edition (0-534-38905-8)

Probability
Bean, *Probability: The Science of Uncertainty with Applications to Investments, Insurance, and Engineering* (0-534-36603-1)

Real Analysis
Gaughan, *Introduction to Analysis, Fifth Edition* (0-534-35177-8)

Paul J. Sally, Jr., Editor
Paul J. Sally, Jr. is the recipient of the American Mathematical Association's 2000 Award for Distinguished Public Service. He was presented this award in recognition of the quality of his research, for his service to the Society as Trustee, and, most importantly, for his many efforts in improvement of mathematics education for the nation's youth (especially for members of minority and underrepresented groups) and for his mentoring of students.

Dr. Sally received his Ph.D. from Brandeis University in 1965. That same year, he joined the faculty at the University of Chicago, where he served as Department Chair from 1977 until 1980. He has founded and guided numerous programs for mathematically gifted youth, in addition to publishing many distinguished research papers, editing several books, and supervising eighteen Ph.D. students.

ABSTRACT ALGEBRA

ABSTRACT ALGEBRA

Ronald Solomon
The Ohio State University

THOMSON
BROOKS/COLE

Australia • Canada • Mexico • Singapore • Spain
United Kingdom • United States

Sponsoring Editor: Bob Pirtle
Assistant Editor: Stacy Green
Editorial Assistant: Jessica Zimmerman
Marketing Manager: Tom Ziolkowski
Marketing Assistant: Stephanie Taylor
Project Manager, Editorial Production: Janet Hill

Print/Media Buyer: Vena Dyer
Production Service: Hearthside Publishing Services
Illustrator: Hearthside Publishing Services
Cover Designer: Jennifer Mackres
Compositor: UG / GGS Information Services, Inc.
Printer: Phoenix Color Corp

COPYRIGHT © 2003 Brooks/Cole, a division of Thomson Learning, Inc. Thomson Learning™ is a trademark used herein under license.

ALL RIGHTS RESERVED. No part of this work covered by the copyright hereon may be reproduced or used in any form or by any means—graphic, electronic, or mechanical, including but not limited to photocopying, recording, taping, Web distribution, information networks, or information storage and retrieval systems—without the written permission of the publisher.

Printed in the United States of America

1 2 3 4 5 6 7 06 05 04 03 02

For more information about our products, contact us at:
Thomson Learning Academic Resource Center
1-800-423-0563

For permission to use material from this text, contact us by:
Phone: 1-800-730-2214
Fax: 1-800-730-2215
Web: http://www.thomsonrights.com

Library of Congress Control Number: 2002113974

Student Edition: ISBN 0-534-39996-7

Brooks/Cole–Thomson Learning
10 Davis Drive
Belmont, CA 94002
USA

Asia
Thomson Learning
5 Shenton Way #01-01
UIC Building
Singapore 068808

Australia/New Zealand
Thomson Learning
102 Dodds Street
Southbank, Victoria
Australia 3006

Canada
Nelson
1120 Birchmount Road
Toronto, Ontario M1K 5G4
Canada

Europe/Middle East/Africa
Thomson Learning
High Holborn House
50/51 Bedford Row
London WC1R 4LR
United Kingdom

Latin America
Thomson Learning
Seneca, 53
Colonia Polanco
11560 Mexico D.F. Mexico

Spain/Portugal
Paraninfo
Calle Magallanes, 25
28015 Madrid
Spain

To Andrea Green

PREFACE

There are numerous excellent textbooks presenting the basics of abstract algebra for college students. Why write another?

Most of the existing texts have a foundational character. They seem designed to lay a firm foundation for future graduate courses in abstract algebra. This is a noble goal and well served by many of the exemplars. There is however a large audience for an undergraduate abstract algobra course consisting of students who will likely never take a graduate course in abstract algebra. Notable among these are the future high school mathematics teachers. These students are better served by a course that emphasizes the roots of abstract algebra, which live in the rich soil of high school mathematics—Euclidean geometry, polynomial algebra, and trigonometry. Out of this soil spring naturally the concepts of symmetry, the complex numbers and the cyclotomic number fields, eventually blossoming into the Galois theory of equations.

The intention of this text is to emphasize the organic and historical development of the abstract theory of groups, rings, and fields from the substrate of high school mathematics. In Part I the "history" is fictitious. It is only with imaginative hindsight that we can attribute the concept of a group of motions to Euclid. In the later parts, however, the history is genuine, although the notation and terminology is updated.

Novel and exciting ideas and theorems are encountered early and often—the 2-dimensional symmetry groups in Chapter 3, Cardano's formulas in Chapter 5, the complex numbers in Chapter 6, the Fundamental Theorem of Algebra in Chapter 7 and the 3-dimensional symmetry groups in Chapter 8. Section 3 on Number Theory features the work of Fermat. Not only his Little Theorem but also the Two Squares and Four Squares Theorems, as well as some cases of the celebrated Last Theorem and its polynomial analogue, are presented. The final Grand Synthesis section begins with a careful treatment of Gauss' proof of the straight-edge-and-compass constructibility of the regular 17-sided polygon, and culminates with Galois' theory of equations. Constructibility of regular polygons is a lovely topic, wonderfully down-to-earth and visual, yet laden with deep connections to subtle topics in number theory and group theory. A course that ends with Chapter 16 (even omitting some of the earlier material) will have presented a rich array of ideas to the students, all closely tied to the most elementary of questions in Euclidean geometry and the study of numbers and polynomial equations. At the end of the Introduction, I discuss some possible syllabi for semester-long and year-long courses.

Typically, Galois Theory appears as the grand finale and raison d'etre for a first course in abstract algebra. But all too often there is not enough time to reach the finale or to give it the attention it deserves. The abstract algebra course then becomes a series of complicated finger exercises with no beautiful sonata to play at the end. Here, too, Galois Theory is the grand finale. But along the way, the students get to play many lovely preludes, nocturnes and sonatinas; so even if the final sonata is never reached, the journey will have been filled with lovely music.

Besides giving a more organic and evolutionary development of the subject, it is the intention of this text to emphasize the connections within algebra and between algebra and other areas of mathematics, especially geometry. The different fields of mathematics are not hermetically sealed off from each other. Quite the contrary, most of the truly important achievements in mathematics have been the product of fruitful interaction of areas.

A textbook is at best a learning aid and at worst a stumbling block. Learning occurs on the dynamic interface between teacher and student. I have had the good fortune to have been inspired by many superb teachers, and wish to acknowledge a few here: Blossom Backal, who taught me high school geometry and first opened my eyes to the beauty of mathematics; Ralph G. Archibald, who taught me number theory and introduced me to mathematical research. David Goldschmidt, who taught me local group theory and gave me glimpses of a truly deep thinker at work; and my thesis advisor, Walter Feit, who taught me representation theory and forced me to figure out for myself why $V = [V, A] \oplus C_V(A)$ is Fitting's Lemma,

With specific regard to this book, I am deeply indebted to Keith Conrad, David Pollack and Inna Korchagina for their comments, corrections and encouragement. Had I saved all of their corrections more systematically, this book would have far fewer errors than it does. Also I and the book owe much to the students I have taught (and learned from) over the years of developing this material. I mention only two representatives of the many who deserve thanks: Jon Spier and Jason Petry.

Ronald Solomon

CONTENTS

0	Background	7

Section One Geometry — 13

1	What Is Congruence?	14
2	Some Two-Dimensional Geometry	22
3	Symmetry	32

Section Two Polynomials — 45

4	The Root of It All	46
5	The Renaissance of Algebra	49
6	Complex Numbers	59
7	Symmetric Polynomials and The Fundamental Theorem of Algebra	68
8	Permutations and Lagrange's Theorem	78
9	Orbits and Cauchy's Formula	86
9A	Hamilton's Quaternions (Optional)	95

Section Three Numbers — 101

10	Back to Euclid	102
11	Euclid's Lemma for Polynomials	113
12	Fermat and the Rebirth of Number Theory	122
13	Lagrange's Theorem Revisited	136
14	Rings and Squares	142
14A	More Rings and More Squares	149
15	Fermat's Last Theorem (for Polynomials)	157
15A	Still more Fermat's Last Theorem (Optional)	165

Section Four The Grand Synthesis — 169

16	Constructible Polygons and the Method of Mr. Gauss	170
17	Cyclotomic Fields and Linear Algebra	177
18	A Lagrange Theorem for Fields and Nonconstructibility	191
19	Galois Fields and the Fundamental Theorem of Algebra Revisited	196
20	Galois' Theory of Equations	207
21	The Galois Correspondence	212
22	Constructible Numbers and Solvable Equations	217

Index — 223

INTRODUCTION

The task of the educator is to make the child's spirit pass again where its forefathers have gone, moving rapidly through certain stages but suppressing none of them. In this regard, the history of science must be our guide.

—Henri Poincaré

In the course of the years 1770 and 1771 Joseph Louis Lagrange, mathematician to the court of Friedrich der Grosse of Prussia, read to the Royal Academy of Sciences and Arts a lengthy memoir entitled: "Réflexions sur la Résolution Algébrique des Équations" (thoughts on the algebraic solution of equations). This memoir did not solve any important problem. It did not explode like a bombshell over the mathematical community of Europe. But it contained an idea. It planted a seed.

Many shells would be fired over Europe and America in the ensuing decades. A new nation, based on a new idea of government, would be founded in America, an old dynasty would come to a bloody end in France, and Europe would be convulsed by 25 years of wars and repeated popular insurrections. New forms of government and industry were being born in violence.

Meanwhile the seed planted by Lagrange continued to germinate in the minds of mathematicians in Italy, Germany, France, and Norway. Some were staunch royalists; some were radical populists. All were assisting at the birth of a new mathematics. Finally on the night of May 29–30, 1832, as the radicals in Paris took to the barricades one more time, a young Frenchman, Évariste Galois, set down his final thoughts on the theory of equations and, in so doing, both laid to rest the problem posed by Lagrange and, much more important, opened the door to a new world of mathematics. In 1951, 119 years later, Hermann Weyl, one of the leaders of mathematics and physics in the early 20th century, would say in a lecture at Princeton University (February 1951):

> Galois' ideas, which for several decades remained a book with seven seals but later exerted a more and more profound influence upon the whole development of mathematics, are contained in a farewell letter written to a friend on the eve of his death, which he met in a silly duel at the age of twenty-one. This letter, if judged by the novelty and profundity of ideas it contains, is perhaps the most substantial piece of writing in the whole literature of mankind.

Great ideas are the enduring legacy of mankind. The empire of Napoléon stretched briefly from Madrid to Moscow and quickly crumbled like the statue of Ozymandias. But the concepts of *liberté, égalité, fraternité*, the proposition that "all men are created equal," and yes, the idea of a group inspire us and shape who we are today.

The principal goal of this course is to trace the history of an idea, from Lagrange to Gauss to Galois. The pursuit of this idea led mathematicians from the concrete world of polynomial equations and regular polygons to the abstract realm of groups, fields, and rings. This brave new world was so rare and strange that 100 years after Galois' death, after the theories of relativity and quantum mechanics had revolutionized physics, after another great European War and a social revolution in Russia, it would still be called *Modern Algebra* in the title of the influential textbook by B. L. Van der Waerden first published in 1931. Like Euclid's *Elements*, Van der Waerden's book represented a summation and codification of a body of knowledge that had been accumulating over the previous 160 years and had finally received an elegant and definitive treatment in the course notes of two great mathematicians of the early 20th century, Emil Artin and Emmy Noether.

Like the layered cities of the ancient Near East, Van der Waerden's algebra quickly subsided into the solid foundation on which a new algebraic edifice would be built out of categories and functors, varieties and morphisms, sheaves and schemes. Thus the adjective *modern* became increasingly inappropriate, and it became fashionable of late for textbooks to dub this material "abstract algebra." "Abstract" is of course a relative term. Any student upon first encountering the "unknown" x in grade school will probably assure you that even this algebra (and hence *all* algebra) is abstract, so the adjective is redundant. At the other extreme, practitioners of late-20th-century algebra will assure you that Van der Waerden's algebra is the concrete foundation for their modern abstractions. All is relative.

Following Van der Waerden, most modern "abstract algebra" texts emphasize the "abstract," beginning like Euclid with a set of *axioms* for a "group" or a "ring" and developing in true Euclidean fashion a collection of theorems about "groups" and "rings," as well as a few examples along the way, since groups and rings are not quite so commonplace as the triangles and circles of Euclid's *Elements*.

The traditional approach has the advantage of efficiency but the defect of severing the material from its historical roots and from its connections with the rest of the body of mathematics. This text attempts a different approach, letting the abstract concepts emerge gradually from less abstract problems about geometry, polynomials, numbers, etc. This is how the subject evolved historically. This is how all good mathematics evolves—abstraction and generalization is forced on us as we attempt to understand the "concrete" and the particular.

In many ways this book is a throwback to a less abstract algebra, driven by the problems that fascinated Lagrange, Gauss, and Abel. They helped fashion a tool—the group—to solve their problem. Galois began the paradigm shift in which the study of the internal structure of the tool itself became more interesting than the solution of the original problem. This is the beginning of abstract algebra and the end of our text. In that sense our text should be called *An Invitation to Abstract Algebra*, and the reader who finds the subject enchanting is advised to study further in texts such as *Topics in Algebra* by Herstein.

Section I begins with Euclid's geometry and anachronistically teases out the concept of a group from his intuitive treatment of congruence. Here symmetry appears in its most visual form with the regular polygons and polyhedra. Later the "hidden symmetries" of the regular polygons will appear in the work of Gauss.

Section II begins the main theme of polynomial equations much as they were studied in Western Europe from the 1500s through the 1700s, and we watch the true historical emergence of the concept of a group as it developed in the conversation of mathematicians in the late 18th and early 19th centuries. It is appropriate here to acknowledge a debt and to recommend enthusiastically the book *Galois' Theory of Algebraic Equations* by Jean-Pierre Tignol, which does a superb and detailed job of tracing how this mathematics was made.

In Section III we turn to the elementary theory of numbers, beginning with Euclid but focusing on its modern reemergence in the cryptic correspondence of Pierre de Fermat and its clarification in the work of Leonhard Euler and Karl Gauss. The concepts of domain and ring emerge as organizing principles and help clarify the similarities and differences between numbers and polynomials. From 1644 to 1994, from Fermat to Andrew Wiles, the Dulcinea of every number theorist's quest was the "Last Theorem" of Fermat. Much of the work of Wiles and its immediate antecedents lies in the domain of still-modern algebra, where this text dares not tread. Nevertheless we do present some older work that helped clarify the important concept of unique factorization.

Finally we return to the subject of polynomial equations and study some of the astonishing work of Karl Gauss on cyclotomic equations and the work of Évariste Galois, which has come to be known as Galois Theory. Here the concept of a field comes to the fore and a remarkable correspondence between fields and groups emerges.

ADVICE TO STUDENTS

A few words are in order about this book as a learning tool. It is likely that most if not all of your previous math textbooks have employed the following format. Each section of the text contains certain illustrative solved problems, designed to serve as templates for most of the exercises. The exercises then, in general, are designed to hone the student's skills at performing the calculations or implementing the algorithms illustrated in the template problems.

Much of mathematics is facilitated by efficient algorithms and the mastery of these algorithms for arithmetic, elementary algebra, calculus, linear algebra, etc., is a foundation on which the study of higher mathematics sits. You have all achieved reasonable mastery at this level. It is time to move on.

The material in this book has relatively little to do with computation and algorithm, and quite a lot to do with concept and theory-building. For this task, template problems are not helpful. What is essential is a careful and critical reading of the text and a precise assimilation of the definitions and concepts.

By and large, the exercises in each section are designed to enable you to build your own understanding of the concepts incrementally. Very often exercises depend heavily on previous exercises. Sometimes you will be advised to use a certain previous exercise, other times not. Usually the exercises begin with fairly easy applications of the definitions and build up gradually.

Don't look for models or templates. You have the resources within yourself to understand the concepts and do the exercises. The material in the first few chapters deals with the elementary Euclidean geometry of the plane and of 6-space. Draw pictures to help yourself visualize what the exercise is saying. Use your common sense. Many of the early exercises are intuitively obvious. Then think about how to translate your common sense into the formal language of mathematics.

At first this may be challenging. Ask questions of your instructor. Work with your classmates.

The level of difficulty of the exercises is uneven. Sometimes an exercise may appear so easy that you will think there is a "trick." Almost certainly there is no trick. It is just a very easy exercise. Very difficult exercises are generally designated as "Bonus Exercises," but some of the regular exercises are also quite challenging.

Abstract algebra has deep and important connections to the other branches of mathematics. Sorry to say (I'm not really sorry), you will be required to remember some basic material that you learned in earlier courses—high school geometry, linear algebra, and even a bit of calculus. Dust off your old textbooks and let's begin!

TO THE INSTRUCTOR

This book is somewhat terse. In consequence I recommend a more leisurely pace than might be suggested by the number of pages. In particular, Chapters 2 and 3 are particularly rich in material and ideas, combining geometry, linear algebra, functions as mappings, and groups. This would certainly be indigestible were it not all grounded in the very concrete visual world of 2-dimensional and 3-dimensional geometry. Nevertheless these two sections warrant a slow and careful treatment. By way of contrast, Chapters 4, 5, and 6 are mostly computational algebra and serve as pleasant "comic relief" after the rich stew of Chapter 3. (Chapter 7 is once again dense with new ideas.) I believe that Chapter 6 deserves careful attention. To my amazement students usually enter this upper-level course with essentially no knowledge of the complex numbers.

At Ohio State, we have two parallel year-long abstract algebra sequences, one designated "honors", the other not. This text is used in the non-honors version and I think of the typical student as a future high school math teacher. For this reason, I have tried to link the material to important themes in high school mathematics—Euclidean geometry, polynomials, numbers, and functions.

In a year-long course, I still do not in general cover all of the material in the book. Indeed, I am satisfied if I get through Chapter 16 and delighted if I get through Chapter 18, while omitting the optional Chapters 9A and 15A.

If you are teaching a one-semester course, I think it is realistic (though I have never tried this) to cover the following material:

Chapters 1–8, 10–12, and 16.

Chapter 7 can be given a "light", heuristic treatment. It may be viewed more as a historical interlude rather than a body of material that needs to be presented rigorously and digested fully. In any case, this set of chapters gives a nice sampling of geometry, polynomial algebra, number theory and group theory, with highlights including Fermat's Little Theorem, Lagrange's Theorem, the Fundamental Theorem of Algebra and Gauss' proof of the constructibility of the regular 17-gon. Rings get short shrift. If your students already know some or all of the basic number theory in Chapters 10–12, then you might want to add Chapters 14 or 15 to touch on some ring theory. Alternatively Chapters 17 and 18 will deepen their grasp of linear algebra and field theory.

It seems impossible to write an algebra text at this level without culminating in the beautiful Galois Theory, and likewise almost impossible to cover this material adequately, except with honors students. I consider it an acceptable loss if future high school math teachers do not see Galois Theory in this course, though of course I cherish the hope that they will be inspired to keep the book and read it themselves some day.

Instead of the usual procedure of placing all of the exercises for a section at the end of the section, I have interwoven the exercises into the body of the text. They are however numbered sequentially throughout each section. I apologize for the extra effort this entails in finding the exercises. As mentioned in the advice to the students above, exercises very often build on previous exercises, even if this is not explicitly indicated in the text. Skipping some exercises is a peril. It may make later otherwise easy exercises difficult, if not impossible, for the students. I have learned that students need to learn that it is not necessary, indeed counterproductive, to keep reinventing the wheel. Theorems in the text and earlier exercises are meant to be used as tools to do later exercises.

Usually, I do one or two of the exercises in class for the students, either before or shortly after they are assigned, by way of illustration or template, especially in the early part of the course. Then I usually give out solutions for most or all of the exercises, after the students have submitted their work.

This material presents two challenges for the students. For me the most important is the intuitive assimilation of the concepts, the internalization of such abstractions as group, ring, and field as almost palpable concrete objects that the students can play with and study. The second is the mastery of the language and syntax of mathematical proof—the translation of intuitive understanding into a formal system which communicates clearly to other mathematicians. These are quite different challenges and the mastery of either in the absence of the other is useless.

CHAPTER 0

BACKGROUND

Although this course will proceed in a somewhat historical path, tracing the evolution of certain basic algebraic ideas such as number, equation, and symmetry from early Greek mathematics to the mid-19th century, we shall from the beginning freely make use of algebraic notation, which did not emerge until the time of Descartes (early 1600s), and of concepts such as set, function, equivalence relation, etc. of an even later vintage. (The term *function* was perhaps first used by Leibniz in the early 1700s.)

Most, if not all, of these concepts should be familiar to you from earlier math courses – precalculus, calculus, linear algebra, and/or foundations of higher mathematics. For this reason, we provide only a very brief review of the most important notations and concepts here. If you need more examples or lengthier explanations, ask your instructor to recommend a good book to consult.

We assume an intuitive notion of the concept *set*. Some important sets for this course are the sets **Z** of integers, **N** of natural numbers (nonnegative integers), **Q** of rational numbers (fractions), **R** of real numbers, and **C** of complex numbers.

Definition. If D and T are two sets, then their Cartesian product is the set

$$D \times T = \{(d, t) : d \in D, t \in T\}.$$

Perhaps the most famous and important Cartesian product is $\mathbf{R} \times \mathbf{R} = \mathbf{R}^2$. If we view $\mathbf{R} = \mathbf{R}^1$ geometrically as the "real number line," then $\mathbf{R}^2$ is the "Cartesian plane."

Absolutely crucial to all that follows is a profound understanding of the concept of a function, so we shall devote some time to this now. There are many ways to think of a function. A function f may be thought of as a rule or correspondence that assigns to each object x in some specified set D a uniquely determined object $f(x)$ in some (possibly different) set T. The set D is then called the **domain** of f and T is sometimes called the **target set** of f. The subset

$$R = \{f(x) \in T : x \in D\}$$

is called the **range** of f. Thus D could be the set of all people living in the United States on January 1, 2001, T could be the set of all colors, and f could assign to each person his or her hair color on January 1, 2001, assuming this could be determined unambiguously.

From this point of view, a function is simply a glorified two-column list, though the set D may be unlistably large. The crucial feature is that the rule that determines $f(x)$ given x must yield the same answer no matter who is applying the rule. In your calculus courses, you have become accustomed to examples of functions with domain $\mathbf{R}$ or $\mathbf{R}^2$ in which the defining rule is given by a formula written in algebraic notation, such as

$$f(x) = x^3 + \frac{4}{x} + 7$$

or

$$f(x, y) = 2^{x+y} + x^2 y - 4.$$

Closely related to a function f is its graph $Gr(f)$, which is the following subset of $D \times T$:

$$Gr(f) = \{(x, f(x)) : x \in D\} \leq D \times T.$$

Clearly f and $Gr(f)$ determine each other uniquely, and some mathematicians, who like to derive all of mathematics from an axiomatic version of set theory, will identify f and $Gr(f)$, avoiding the undefined term *rule* or *correspondence*.

These ways of thinking about functions, though logically equivalent to all others, are rather static. We shall prefer a dynamic concept of a function as a **map** or **mapping**. We shall often wish to think of f as moving the "point" P in D to the point $f(P)$ in T. Indeed usually in this course, the sets D and T will coincide, and we shall think of f as a "deformation" of D or as a "reshuffling" or **permutation** of the points of D. This dynamic view is implicit in the notation $f : D \to T$.

For example, we may think of the function $f : \mathbf{R} \to \mathbf{R}$ given by the rule

$$f(x) = x^2$$

as a rule for folding the real number line $\mathbf{R} = \mathbf{R}^1$ onto its nonnegative half, "pinning" it together at 0 and 1, stretching out the subset $(1, \infty)$ and squeezing the points of $(0, 1)$ closer toward 0.

Likewise we may think of the function $r : \mathbf{R}^2 \to \mathbf{R}^2$ given by the rule

$$r(x, y) = (-y, x)$$

as a rule for rotating the Euclidean plane $\mathbf{R}^2$ 90° counterclockwise about the point $(0, 0)$.

With this dynamic terminology, we shall speak of a point x in the domain of the function f as a **fixed point** of f or as **fixed** by f if $f(x) = x$. Thus in the first example, the fixed points of f are the points 0 and 1, and in the second example, the unique fixed point of r is the point $(0, 0)$.

Exercises

0.1. Consider the function $T : \mathbf{R}^2 \to \mathbf{R}^2$ defined by the rule $T(x, y) = (-x, y)$. Describe T geometrically as a mapping of the plane. What are the fixed points of T?

0.2. Consider the function $S : \mathbf{R}^2 \to \mathbf{R}^2$ defined by the rule $S(x, y) = (x^2, y)$. Describe S as a mapping of the plane and determine the fixed points of S. ∎

We shall be particularly interested in invertible functions. The following definitions are relevant in this context.

Definition. We say that f is an **onto** or **surjective** function if $R = T$. For each $r \in R$, we set
$$f^{-1}(r) = \{d \in D : f(d) = r\}.$$
We say that f is a **one-to-one** or **injective** function if $f^{-1}(r)$ contains only one element for each $r \in R$. If f is both one-to-one and onto, then we say that f is **bijective** or a **one-to-one correspondence** or an **isomorphism** between D and T. If $D = T$ and $f : D \to D$ is bijective, then often f is called a **permutation** of the set D, especially if D is a finite set. Whenever $f : D \to T$ is bijective, the set
$$\{(t, f^{-1}(t)) : t \in T\} \leq T \times D$$
is the graph of a function $f^{-1} : T \to D$. We call f^{-1} the **inverse** of the function f.

Definition. We say that the sets D and T have the same **cardinality** if there is a one-to-one correspondence $f : D \to T$. We write $|D|$ to denote the cardinality of the set D. If D has the same cardinality as the set of natural numbers $\{1, 2, \ldots, n\}$, then we write $|D| = n$.

Pigeon-Hole Principle. *If A is a subset of the finite set B and if A and B have the same cardinality, then $A = B$.*

Note that the Pigeon-Hole Principle is true *only* for finite sets. For example the set **N** of natural numbers and the set **2N** of even natural numbers have the same cardinality but are obviously not equal.

If $D = T$, then we can form the **composition** $f \circ g$ of two functions $f : D \to D$ and $g : D \to D$:
$$f \circ g(d) = f(g(d)).$$
Of course we can also form $g \circ f$, and usually these two functions are not equal. Again it is useful to think of $f \circ g$ dynamically as "first do g to D, then do f." This operation of composition of functions will be of absolutely crucial significance to us.

For any set D, there is a distinguished bijective function $I : D \to D$, called the **identity function**, defined by the rule
$$I(x) = x$$
for all $x \in D$. Thus, speaking dynamically, I is the function that "does nothing" to D. Every point of D is a fixed point of I.

If $f : D \to D$ is a bijective function, then the inverse function f^{-1} is characterized by the functional equation
$$f \circ f^{-1} = I = f^{-1} \circ f.$$
Note that in all cases, equality of functions
$$f = g$$
means
$$f(x) = g(x) \text{ for all } x \in D,$$
where D is the domain of both f and g.

We continue this background section with a brief discussion of the concept of a relation on a set D, with particular attention to equivalence relations.

Definition. A **relation** R on a set D is any subset of $D \times D$. If $(a, b) \in R$, we say that a is in the relation R to b.

R is said to be **reflexive** if R contains the diagonal set

$$\Delta = \{(d, d) : d \in D\}.$$

R is **symmetric** if the following holds:

$$\text{If } (a, b) \in R, \text{ then } (b, a) \in R.$$

R is **transitive** if the following holds:

$$\text{If } (a, b) \in R \text{ and } (b, c) \in R, \text{ then } (a, c) \in R.$$

Finally we call R an equivalence relation if R is reflexive, symmetric, and transitive.

If E is an equivalence relation on D and if $a \in D$, then the set

$$E_a = \{d \in D : (a, d) \in E\}$$

is called the (E)-**equivalence class** containing a. Obviously every element of D lies in exactly one E-equivalence class. Thus the set of E-equivalence classes forms a **partition** of D into disjoint subsets. Conversely any partition P of D into disjoint subsets defines an equivalence relation E_P via $(a, b) \in E_P$ if and only if a and b lie in the same subset of the partition P.

Equivalence relations are central to mathematics. The quintessential equivalence relation is equality, denoted by $=$. It is the finest of equivalence relations: An object is equal only to itself. All other equivalence relations may be thought of as "filters" through which we look at mathematical objects, ignoring certain differences and allowing only the similarities to show through. Thus in Euclidean geometry, the congruence filter ignores the position of objects and observes only their size and shape, while the even darker similarity filter also ignores size and observes only shape. Equivalence relations are crucial to our ability to make generalizations, to prove general theorems.

At a deeper level, an equivalence relation on one mathematical entity may be used to define a new mathematical entity whose members are the equivalence classes themselves. We shall see some examples of this type of construction, which appeared first in the work of Gauss around 1800 and was one of the key constituents of the newly emerging abstract algebra.

We shall repeatedly be considering sets with "operations" in the following sense.

Definition. A set D is said to be **closed** under a binary operation $\times$ (sometimes denoted $+$ or $\circ$, etc.) if $\times$ is a function:

$$\times : D \times D \to D.$$

Usually we denote $\times(a, b)$ by $a \times b$. We say that $\times$ is an **associative operation** if

$$a \times (b \times c) = (a \times b) \times c$$

for all $a, b, c \in D$, and we say that it is a **commutative operation** if

$$a \times b = b \times a$$

for all $a, b \in D$.

Finally let's tie functions, equivalence relations, and sets with operations together. Often an equivalence relation is defined by a collection $\mathcal{F}$ of bijective functions such that two objects A and B are said to be equivalent (or more frequently, **isomorphic**) if there is a function $f \in \mathcal{F}$ with $f(A) = B$. When the objects A and B are sets with operations (so that we have $(A, \times)$ and $(B, +)$, for example), then we will require the function f to satisfy

$$f(a \times a') = f(a) + f(a') \text{ for all } a, a' \in A.$$

Some of this may be bewildering. Don't worry. It is the purpose of this course to provide examples that shed light on these abstractions.

The material of calculus beyond the functional concept will play very little role in this course, although the derivative of a polynomial will surface occasionally. A larger role will be played by linear algebra. Many of the fundamental concepts will be reviewed when needed. A basic familiarity with matrix algebra (addition and multiplication) and the elementary properties of the trace and determinant of a matrix will be assumed.

Exercises

0.3. Make up an example of a relation that is symmetric and transitive but not reflexive. Make up another that is reflexive and transitive but not symmetric. Make up a third that is reflexive and symmetric but not transitive. ∎

Occasionally proofs in the text, and proofs requested of you as exercises, will best be done by Mathematical Induction. Sometimes students come to believe that Mathematical Induction is a "silver bullet" which should be used for all proofs. This is far from the case. Rarely is "induction" the appropriate tool in any of the exercises. When it is, you will usually be advised to use it.

The Principle of Mathematical Induction. *Suppose that $\{P(0), P(1), \ldots\}$ is a set of mathematical statements indexed by the natural numbers. Suppose further that the following two statements are true:*

(a) The statement $P(0)$ is true.
(b) For any natural number n, the truth of $P(n)$ implies the truth of $P(n+1)$.

Then all of the statements $P(0), P(1), \ldots$ are true.
In fact this is really just a statement about sets of natural numbers:
Let S be a set of natural numbers about which the following statements are true:

(a) 0 is in S.
(b) If n is in S, then also $n + 1$ is in S.

Then S is the set $\mathbf{N}$ of all natural numbers.

An equivalent statement, the Well-Ordering Principle for **N**, will be formulated later in the text.

Another equivalent statement is the following.

The Principle of Complete Mathematical Induction. *Suppose that* $\{P(0), P(1), \ldots\}$ *is a set of mathematical statements indexed by the natural numbers. Suppose further that the following two statements are true:*

(a) $P(0)$ is true;

(b) For any natural number n, the truth of $P(k)$ for all k less than n implies the truth of $P(n)$.

Then all of the statements $P(0), P(1), \ldots$ are true.

You will see some examples of proofs by (complete) mathematical induction in the text, and you will have some opportunities in the exercises to practice creating such proofs yourself.

SECTION ONE
GEOMETRY

CHAPTER 1   WHAT IS CONGRUENCE?

CHAPTER 2   SOME 2-DIMENSIONAL GEOMETRY

CHAPTER 3   SYMMETRY

CHAPTER 1

WHAT IS CONGRUENCE?

Classical mathematics is a quest for structural harmony. It began with the realization by ancient Greek geometers that our 3-dimensional continuum possessed a remarkable symmetry which permeates the essential properties of the physical world.

– Mikhael Gromov (1998)

Geometry is not the main focus of this book. Nevertheless during the period that gave birth to abstract algebra, questions about the foundations of geometry were an important recurring topic. Near the beginning of the period (around 1800) came Gauss' realization of the possibility of a logically consistent "non-Euclidean" geometry, followed by independent similar realizations by Bolyai and Lobachevski. In 1827, just before Galois' fundamental work on groups and the theory of equations, Gauss' student, Möbius, wrote a paper studying certain motions of the plane preserving various types of geometrical properties: straightness of lines, length, shape, parallelism. Möbius considered compositions of motions but did not quite grasp the group concept. Later came two famous documents, Riemann's inaugural lecture at Göttingen on the foundations of geometry (1854) and then Felix Klein's "Erlanger Programm" (1872), in which the ideas implicit in Möbius' paper were finally clearly articulated. We shall develop some of these ideas here, aiming in particular at a mathematical formulation of symmetry.

Symmetry (especially bilateral symmetry) has been a source of fascination throughout recorded history. The earliest geometry "theorems," attributed to Thales (c. 585 B.C.), feature symmetrical images:

1. The vertical angles formed by two intersecting lines are equal.
2. The base angles of an isosceles triangle are equal.
3. A circle is bisected by any line through its center.

Geometry is concerned with "figures" composed of "points" and their comparison in terms of size, shape, and position. We are primarily interested in the relation between figures called **congruence**, which holds between two objects of the same "size and shape." How can we verify the congruence of two figures? Euclid's fundamental congruence axiom is the following as given in the translation by Sir Thomas Heath.

> Common Notion 4. Things which coincide with one another are equal to one another.

Following standard modern usage it would be preferable to use the word *congruent* in place of "equal." But still, what does Euclid mean by this? We quote Heath's commentary:

> It seems clear that the Common Notion, as here formulated, is intended to assert that superposition is a legitimate way of proving the equality of two figures which have the necessary parts respectively equal, or, in other words to serve as an axiom of congruence.
>
> The phraseology of the propositions, e.g. I.4 and I.8, in which Euclid employs the method indicated, leaves no room for doubt that he regarded one figure as actually moved and placed upon the other.

Let us look at the famous and important Proposition 4 of Book I.

Side-Angle-Side Proposition. If two triangles have the two sides equal respectively to two sides respectively, and have the angles contained by the equal straight lines equal, they will also have the base equal to the base, the triangle will be equal (congruent) to the triangle, and the remaining angles will be equal to the remaining angles respectively, namely those which the equal sides subtend.

Here is a slightly condensed version of Euclid's proof as given by Heath:

Proof. Let ABC, DEF be two triangles having the two sides AB, AC equal to the two sides DE, DF respectively, and the angle BAC equal to the angle EDF. If the triangle ABC be "applied" to the triangle DEF, and if the point A be placed on the point D and the straight line AB on DE, then the point B will also coincide with E because AB is equal to DE.

Again, AB coinciding with DE, the straight line AC will also coincide with DF, because the angle BAC is equal to the angle EDF. Hence the point C will also coincide with the point F, because AC is again equal to DF. But B also coincided with E; hence the base BC will coincide with the base EF. Thus the whole triangle ABC will coincide with the whole triangle DEF, and will be equal (congruent) to it. ∎

Let us reconsider Euclid's method from a more modern viewpoint. Clearly Euclid wishes to partition the collection of all geometrical figures (or at least all triangles) into disjoint sets (or congruence classes). So congruence in geometry should be an equivalence relation on the collection of all geometric figures. But we still don't know how to define it. Euclid implies that it should be defined in terms of a process of moving one figure and placing it on top of another figure. How can we make this mathematically precise?

The mathematics of "dynamical processes" such as "moving" and "placing" is tied up with the fundamental concept of a function. There is evidence that the Greek geometers (starting with Plato, and with the notable exception of Archimedes) were uncomfortable with dynamics. Certainly the concept of a function emerged very slowly in the history of mathematics. The term appears to have been used first by Leibniz in the early 1700s.

Nevertheless with the benefit of 20/20 hindsight we wish to make precise Euclid's notion of congruence using a family of functions that map (move) the set P of all points in our geometry onto itself.

Definition. A surjective function $f : P \to P$ is called an **isometry** if f "preserves" distance; i.e.,

$$d(f(p), f(q)) = d(p, q)$$

for all points p and q in P, where $d : P \times P \to \mathbf{R}$ is the distance function on P.

(For many spaces, in particular for Euclidean spaces, it is possible to prove that a distance-preserving map has to be an onto function. The proof is a bit difficult, however, and we choose to make it part of the definition of an isometry.)

Here we need very few assumptions about the distance function, only that d assigns to each pair (p, q) of points a nonnegative real number with the properties that

(a) $d(p, q) = 0$ if and only if $p = q$, and
(b) $d(p, q) = d(q, p)$.

Although it will play no role in the theorems or exercises that follow, we also record the third fundamental property of a distance function, namely the Triangle Inequality:

(c) $d(p, r) \leq d(p, q) + d(q, r)$ for all p, q, r in P.

Let's use these definitions to prove our first little theorem.

Theorem. *Every isometry $f : P \to P$ is a one-to-one function, hence a bijective function.*

Proof. Let p and q be two points in P for which $f(p) = f(q)$. Then

$$d(p, q) = d(f(p), f(q)) = 0,$$

and so $p = q$. Thus f is a one-to-one function. ∎

Note how we have reasoned. If f were not one-to-one, then there would be two different points p and q that f maps to the same point $f(p) = f(q)$. Then the distance between $f(p)$ and $f(q)$ is 0, since they are the same point. But f is distance preserving. So p and q must have also been at distance 0 from each other; in other words, p and q were not different after all. Hence f is one-to-one.

After we have thought through the proof, we can find a concise and elegant way to write it down. Much of our taste in mathematical exposition was formed by the great master, Karl Gauss, who said, "When an architect completes a building, he does not leave the scaffolding standing." You don't have to be great architects yet, but it's time to start building a few buildings of your own in the exercises that follow. Actually you will be constructing a rather large and impressive edifice one small unit at a time. Many of the exercises in this section and throughout the book build on exercises you have done before. That is the way mathematics works. Don't just "do" the exercises. Understand them and expect to use them in your solution of later exercises.

In these first two exercises you will prove two fundamental properties of all isometries. You may be uncomfortable at first with the idea of proving facts about a function f, when you don't know what f is but only know one attribute of f. Rejoice!! The fewer facts you know about f, the fewer facts you can use to prove the exercise. This should focus your thinking.

Exercises

1.1. Prove that if $f : P \to P$ is an isometry, then $f^{-1} : P \to P$ is also an isometry.

1.2. Prove that if $f : P \to P$ and $g : P \to P$ are isometries, then $f \circ g$ is also an isometry.

Thinking of a geometrical figure F as a set of points, we see that f maps figures to figures.

Definition. Two figures F and F' are **congruent** (and we write $F \cong F'$) if there is an isometry f such that $f(F) = F'$.

Let's consider a (too) elementary example to get warmed up. ∎

Example The one-dimensional "Euclidean geometry" where the points are the set of all points on the real number line $\mathbf{R}^1$.

We may identify each point with a real number. The standard Euclidean definition of distance is
$$d(a, b) = |b - a|.$$

Exercises

1.3. Let $\mathbf{R}^1$ be the real number line with the Euclidean distance function d as just defined. Prove: If a and b are distinct points of $\mathbf{R}^1$, there is a unique point c equidistant from a and b. ∎

Let's determine all isometries of $\mathbf{R}^1$. The next exercise is just to refresh your memory from calculus. It is not needed later.

1.4. Prove: If $f : \mathbf{R}^1 \to \mathbf{R}^1$ is an isometry, then f is a continuous function.

The most obvious isometry is the identity function $I : \mathbf{R}^1 \to \mathbf{R}^1$ given by $I(a) = a$ for all $a \in \mathbf{R}^1$.

1.5. Prove: If $f : \mathbf{R}^1 \to \mathbf{R}^1$ is an isometry that fixes two distinct points a and b, then $f = I$. (Suppose on the contrary that there exists $t \neq f(t)$. Use Exercise 1.3.) ∎

Now let's try to think of some more examples of isometries. One way to preserve distance is simply to slide everything a fixed distance to the left or right. Such a map is called a **translation**. For each real number a we have a translation T_a defined by

$$T_a(x) = a + x \text{ for all points } x \text{ on } L.$$

Exercises

1.6a. Prove: For each number a, the map $T_a : \mathbf{R}^1 \to \mathbf{R}^1$ is an isometry.

1.6b. What is T_a^{-1}?

1.7. Prove: If $f : \mathbf{R}^1 \to \mathbf{R}^1$ is an isometry with $f(0) = a$ and $f(1) = a + 1$, then $f = T_a$. (*Hint*: Consider the isometry (why?) $T_a^{-1} \circ f$ and apply Exercise 1.5. This is a basic trick of the trade.)

1.8. Prove: There are exactly two isometries I and R of $\mathbf{R}^1$ mapping 0 to 0. (*Hint*: First figure out what the formula for R is. Then verify that R is an isometry. Now let $T : \mathbf{R}^1 \to \mathbf{R}^1$ be any isometry that satisfies $T(0) = 0$. Use the same trick as in Exercise 1.7.) ∎

Now we can describe all isometries of $\mathbf{R}^1$.

Theorem. *Let $f : \mathbf{R}^1 \to \mathbf{R}^1$ be an isometry. Let $f(0) = a$. Then either $f = T_a$ or $f = T_a \circ R$, where R is the isometry defined in Exercise 1.8.*

Proof. Let $g = T_{-a} \circ f$. By Exercises 1.2 and 1.6, g is an isometry of $\mathbf{R}^1$. Also

$$g(0) = (T_{-a} \circ f)(0) = T_{-a}(f(0)) = T_{-a}(a) = a - a = 0.$$

Thus by Exercise 1.8 either $g = I$ or $g = R$.
Suppose that $g = I$. Then

$$T_{-a} \circ f = I$$

and so

$$T_a \circ (T_{-a} \circ f) = T_a \circ I = T_a.$$

But

$$T_a \circ (T_{-a} \circ f) = (T_a \circ T_{-a}) \circ f = I \circ f = f$$

by the Associative Law for the composition of functions. Hence $f = T_a$ in this case.
Similarly if $g = R$, then

$$f = (T_a \circ T_{-a}) \circ f = T_a \circ (T_{-a} \circ f) = T_a \circ g = T_a \circ R,$$

completing the proof. ∎

Exercises

1.9. Denote by R_a the isometry $T_a \circ R$. Give a formula for R_a. Describe R_a geometrically. What is R_a^{-1}?

Notice that the isometry $I = T_0$ and the isometry $R = R_0$. Also notice that we have now proved that every isometry of $\mathbf{R}^1$ is not only a continuous function, it is even a linear function.

1.10. Is $T_a \circ R = R \circ T_a$? If not, what is a correct formula? Use this to simplify

$$(T_a \circ R) \circ (T_b \circ R).$$ ■

Now let's return to general isometries and consider how we might prove the following assertion.

Theorem. *Let P be a set of points on which a distance function is defined. Congruence of figures is an equivalence relation.*

Proof. We need to establish the three properties of an equivalence relation. Let F, F', and F'' be any three figures.

(a) $F \cong F$.

We must produce an isometry of P that maps F onto itself. The obvious choice is the "identity map" I defined by

$$I(x) = x \text{ for all } x \in P.$$

Since I maps every point to itself, it maps every figure F onto itself. So (a) holds.

(b) If $F \cong F'$, then $F' \cong F$.

The hypothesis that $F \cong F'$ means that there is an isometry $f : P \to P$ with the property that $f(F) = F'$. To verify the conclusion, we must produce an "inverse" isometry mapping F' back to F. Hence we take f^{-1}, which we have shown to be an isometry in Exercise 1.1 and which by definition maps F' onto F.

(c) If $F \cong F'$ and $F' \cong F''$, then $F \cong F''$.

The hypotheses means that there are two isometries $f : P \to P$ and $g : P \to P$ such that $f(F) = F'$ and $g(F') = F''$. Then by the definition of composition of functions, we have $g \circ f(F) = F''$ and we are done, since we have verified in Exercise 1.2 that $g \circ f$ is an isometry. ■

We now arrive at our *main point*. The verification that congruence of figures on P is an equivalence relation practically forced us to verify the following three statements.

(GF1) The composition $f \circ g$ of any two isometries f and g is itself an isometry.

(GF2) The identity function $I : P \to P$ defined by $I(x) = x$ is an isometry.

(GF3) The inverse f^{-1} of an isometry f exists and is itself an isometry.

These conditions on a set of invertible functions have proven to be so fundamental that they have been formalized as the definition of a **group (of functions)**. (The term *group* was first used in this sense by Galois in 1832. A somewhat more formal definition of this nature was made by Cauchy in a paper of 1844 in which he discussed "**composable systems of permutations**".)

Definition. A set G (of functions), on which a binary operation $\cdot$ (composition or multiplication) is defined, is called a **group (of functions)** if the following conditions hold:

(G1) (Closure) If $f \in G$ and $g \in G$, then $f \cdot g \in G$.

(G2) (Identity) There exists an element $e \in G$ with
$$e \cdot f = f = f \cdot e \text{ for all } f \in G.$$

(G3) (Inverses) For each $f \in G$, there exists $f^{-1} \in G$ satisfying
$$f^{-1} \cdot f = e = f \cdot f^{-1}.$$

(G4) (Associative Law) For all $f, g, h \in G$,
$$f \cdot (g \cdot h) = (f \cdot g) \cdot h.$$

When the objects in G are functions and the operation is composition of functions, then the Associative Law (G4) is an immediate consequence of the definition of composition of functions. However, later we shall see examples of groups whose **elements** are numbers or other types of objects. Thus it is useful to have a general definition of an **abstract group** that covers all cases and in which the Associative Law is clearly stipulated as an axiom. For this reason too we use the "neutral" symbol $\cdot$ for the binary operation on G. When we *know* what group (or kind of group) we are talking about, we will sometimes prefer to use the symbol $\circ$ (for functions) or $+$ (for numbers) in place of $\cdot$. Often in fact we will use no symbol at all and simply write gh as the "product" of the elements g and h.

Notice that we have characterized the identity element of G by its relationship to the other elements of G, not by its action on some set P.

Exercises

Let G be a group of bijective functions on a set P.

1.11a. Prove: If $e : P \to P$ is a function in G satisfying
$$e \circ f = f \text{ for all } f \in G,$$
then $e = I$, the identity function on P.

1.11b. Prove: If $e : P \to P$ is a function in G satisfying
$$f \circ e = f \text{ for all } f \in G,$$
then $e = I$. ■

Now we return to geometry. Our result concerning congruence of figures being an equivalence relation is roughly equivalent to the following result.

Theorem. *Let P be a set of points. Let $Isom(P)$ denote the set of all isometries of P. Then $Isom(P)$ is a group of functions on P (called the **isometry group** of P).*

Proof. Since composition of functions is always an associative operation, Exercise 1.2 yields that $Isom(P)$ is closed under the associative operation of composition of isometries. Thus (G1) and (G4) hold. Since the identity function I on P is clearly an isometry, it is in

$Isom(P)$ and satisfies the identity condition $f \circ I = f = I \circ f$ for all $f \in Isom(P)$. Hence (G2) holds. Finally Exercise 1.1 asserts that the inverse of an isometry f of P is an isometry f^{-1} of P and satisfies the inverse condition $f^{-1} \circ f = I = f \circ f^{-1}$. Thus (G3) holds and $Isom(P)$ is a group. ∎

A confession: This is not a book about geometry. Our main motivation in this section was to sneak up on the fundamental concept of a group. An important closely related concept is the following.

Definition. A subset H of a group $(G, \cdot)$ is called a **subgroup** of G if $(H, \cdot)$ is itself a group; i.e., H is a nonempty subset of G and whenever f and g are in H, then so are $f \cdot g$ and f^{-1}.

Exercises

Let $E(1) = Isom(\mathbf{R}^1)$ be the group of all isometries of the Euclidean line (as studied earlier).

1.12. Prove: The set T of all translations of $\mathbf{R}^1$ is a subgroup of $E(1)$.

1.13a. Prove: The Commutative Law holds in the subgroup T; i.e.,
$$T_a \circ T_b = T_b \circ T_a \text{ for all } a, b \in \mathbf{R}^1.$$

1.13b. Prove: The Commutative Law does not hold in $E(1)$. ∎

(A group in which the Commutative Law holds is called an **abelian group** in honor of the mathematician Nils Henrik Abel, who first studied them. A group in which the Commutative Law does not hold is called a **nonabelian group**. Thus T is an abelian group, but $E(1)$ is a nonabelian group.)

A Few Abstract Group Exercises

1.14. Prove: Every subgroup of an abelian group is also an abelian group. However a subgroup of a nonabelian group may be either abelian or nonabelian.

1.15. Prove: Let $(G, \cdot)$ be a group with identity element e. Suppose that $a \in G$ with
$$a \cdot g = g \text{ for all } g \in G.$$
Then $a = e$. (In particular, the identity element e is unique.)

1.16. Prove that the Cancellation Laws hold in groups; i.e., let $(G, \cdot)$ be a group. If g, h, k are elements of G with $g \cdot h = g \cdot k$, then $h = k$. Also if $g \cdot h = k \cdot h$, then $g = k$.

1.17. Prove: Let $(G, \cdot)$ be a group with identity element e. Suppose then $f, g \in G$ with $f \cdot g = e$. Then $f = g^{-1}$. (In particular, the inverse of an element g is unique.)

1.18. Prove: Let $(G, \cdot)$ be a group with identity element e. Suppose that $g \cdot g = e$ for all $g \in G$. Then G is an abelian group. ∎

CHAPTER 2

SOME TWO-DIMENSIONAL GEOMETRY

Greek geometry is early but in a sense completely modern mathematics. It does however show more clearly than later mathematics two peculiarities of the human brain that produced it:

> *(1) it uses the human visual system, in fact geometry is directly derived from visual experience and intuition.*
>
> *(2) it uses an external memory in the form of a drawing formed of lines and circles, with points labelled by letters.*

Combining these two tricks permits elaborate logical constructions which the Greeks rightly considered as prodigious intellectual feats. Hilbert's version of Euclidean geometry without the help of (1) and (2) shows how hard the subject really is.

—David Ruelle (1998)

Although we have elucidated the concept of congruence, we still haven't completely solved Euclid's "problem": how to give a rigorous proof of the SAS Theorem. For this we need to study the isometry group of two-dimensional Euclidean geometry.

Two-dimensional Euclidean geometry is a very rich source of interesting mathematics, and we shall only be able to touch on some of it. Since we are interested in doing algebra, we shall follow Descartes and Fermat in introducing "Cartesian coordinates" in the Euclidean plane. So we shall, as usual, treat the plane as

$$\mathbf{R}^2 = \{(x, y) : x, y \in \mathbf{R}\}.$$

Also we shall make the plane Euclidean by using the Euclidean distance function d defined by

If $u = (x, y)$ and $v = (x', y')$, then $d(u, v) = ((x - x')^2 + (y - y')^2)^{1/2}$.

Notice that this is equivalent to assuming the validity of the Pythagorean Theorem in our geometry. Obviously following Euclid (and Hilbert) we could start from other axioms and prove the Pythagorean Theorem, but we shall start here.

Exercise

2.1. Let $\overrightarrow{OA}$ and $\overrightarrow{OB}$ be the rays emanating from the point O passing through A and B respectively. Let us define the **angle** formed by these rays to be the pairs of supplementary angles (in the usual sense) formed by the rays. Using the Law of Cosines argue that a Euclidean isometry "preserves angles"; i.e., if f is a Euclidean isometry mapping O to O', A to A', and B to B', then the angle formed by the rays $\overrightarrow{OA}$ and $\overrightarrow{OB}$ has the same measure as the angle formed by the rays $\overrightarrow{O'A'}$ and $\overrightarrow{O'B'}$. ■

We want to find all isometries of plane Euclidean geometry. What kind of function is an isometry of $\mathbf{R}^2$? First of all it is a function whose domain and range is $\mathbf{R}^2$, so in the language of calculus it is a vector-valued function of two variables (where the vectors also have two entries). Any such function can be written as

$$f(x, y) = (f_1(x, y), f_2(x, y))$$

where f_1 and f_2 are "scalar-valued" functions of two variables; i.e., their range is contained in the ordinary real numbers. Again we have

Exercise

2.2. Prove: If $f : \mathbf{R}^2 \to \mathbf{R}^2$ is an isometry, then f is a continuous function. ■

OK. Now let's think of some examples that are isometries. As in the one-dimensional case, we have the **translation** maps. If $u = (a, b)$ is a point in $\mathbf{R}^2$, then we define the translation map as

$$T_u(x, y) = (x + a, y + b).$$

Exercises

2.3. Prove that T_u is an isometry of the Euclidean plane for every $u = (a, b)$.

2.4. Prove: If u is not the **0** vector, then T_u has no fixed points; i.e., there is no point $v = (c, d)$ such that $T_u(v) = v$. (On the other hand, what is T_0?)

2.5. Prove that the set $T = \{T_u : u \in \mathbf{R}^2\}$ is an abelian group of isometries of $\mathbf{R}^2$.

2.6. Prove that if u and v are any two points in $\mathbf{R}^2$, there exists a translation T_w in T with $T_w(u) = v$. ∎

Factorization Theorem. *If S is any isometry of $\mathbf{R}^2$, then there is a translation T_w in $\mathbf{T}$ such that $S = T_w \circ R$ for some isometry R of $\mathbf{R}^2$ with $R(0, 0) = (0, 0)$.*

Proof. Let $w = S(0, 0)$. Let $R = T_{-w} \circ S$. Then R is an isometry by Exercises 1.2 and 2.3. Also

$$R(0, 0) = T_{-w} \circ S(0, 0) = T_{-w}(S(0, 0)) = T_{-w}(w) = w - w = (0, 0).$$ ∎

This theorem is very important. It gives a "factorization" of a general isometry into a product of two "simpler" isometries, just like the usual factorization of numbers into products of prime numbers. This is a very important principle in algebra: Try to decompose complex things as products of simple things. Then try to understand the simple things.

But in what sense are the factor isometries simpler? Well, we understand translation maps very well. Let us next try to understand the isometries of $\mathbf{R}^2$ that fix the point $(0, 0)$. As in the one-dimensional case, there are reflections. Although in the one-dimensional case there is only one reflection fixing 0, now we can take the mirror reflection across *any* line passing through $(0, 0)$.

Exercises

2.7a. Write the formula for the reflection map across the x-axis.

2.7b. Write the formula for the reflection map across the y-axis.

Now consider a line ℓ passing through $(0, 0)$ and through $u = (\cos\frac{\theta}{2}, \sin\frac{\theta}{2})$, which is not one of the coordinate axes. Draw a picture showing ℓ, the vector from $(0, 0)$ to $(1, 0)$, and the vector from $(0, 0)$ to the point (c', d') that is the mirror image of $(1, 0)$ across the line ℓ. (For visual clarity, it is easiest when $0 < \theta < \pi$.)

2.8a. By elementary trigonometry, write a formula (in terms of θ) for the coordinates of the point of intersection of ℓ and the line joining $(1, 0)$ and (c', d'). (*Note*: This point is a multiple of u. What factor should you multiply by?)

2.8b. Using 2.8a, show that $(c', d') = (\cos\theta, \sin\theta)$. (Convince yourself that this is valid even if θ is an obtuse angle.)

2.8c. Now find a formula for the mirror image of $(0, 1)$ across ℓ.

2.9a. Let ℓ^* be the line whose equation is $y = mx + b$. Let (c, d) be any point in the plane. Using elementary algebra and geometry, find a formula in terms of m, b, c, and d for the coordinates of the mirror image of (c, d) across the line ℓ^*.

2.9b. Write the special case of the formula from 2.9a when $b = 0$ and $m = \tan\frac{\theta}{2}$. Verify that it agrees with the results of Exercise 2.8 when $(c, d) = (1, 0)$ or $(0, 1)$. ∎

Now let r_ℓ denote the reflection across the line ℓ from Exercise 2.8. We recall the definition of a linear transformation mapping $\mathbf{R}^n$ to $\mathbf{R}^n$.

Definition. A function $T : \mathbf{R}^n \to \mathbf{R}^n$ is a linear tranformation of $\mathbf{R}^n$ if
$$T(au + bv) = aT(u) + bT(v)$$
for all real numbers a, b and all vectors u, v in $\mathbf{R}^n$.

Exercises

2.10. Prove: $T : \mathbf{R}^2 \to \mathbf{R}^2$ is a linear transformation if and only if there exist real numbers $a, b, c,$ and d such that
$$T(x, y) = (ax + by, cx + dy) \text{ for all } (x, y) \in \mathbf{R}^2.$$

Note: For many computational purposes it is more convenient to write the vectors of $\mathbf{R}^2$ as column vectors: $u = \begin{bmatrix} x \\ y \end{bmatrix}$, in which case the formula in 2.10 may be written in matrix form as
$$T\begin{bmatrix} x \\ y \end{bmatrix} = \begin{bmatrix} a & b \\ c & d \end{bmatrix}\begin{bmatrix} x \\ y \end{bmatrix}.$$

We call
$$\begin{bmatrix} a & b \\ c & d \end{bmatrix}.$$
the matrix associated to the linear transformation T. (Strictly speaking, we should say that it is the matrix associated to the linear transformation T with respect to the basis $\{(1, 0), (0, 1)\}$ for $\mathbf{R}^2$. We shall not, however, have occasion to worry about change of basis in this course, and so we shall usually sloppily suppress all reference to the choice of basis.)

2.11. Prove: The reflection r_ℓ is a linear transformation of $\mathbf{R}^2$, and its associated matrix is R_ℓ:
$$\begin{bmatrix} \cos\theta & \sin\theta \\ \sin\theta & -\cos\theta \end{bmatrix}$$

2.12. Let R_x and R_y be the reflections in Exercises 2.7a and 2.7b respectively. Consider the function $\rho = R_x \circ R_y$. Is ρ a reflection? Is ρ an isometry fixing $(0, 0)$. What is ρ? ∎

From Exercise 2.12, we see that not all isometries fixing $(0, 0)$ are reflections. There is another important class of isometries of $\mathbf{R}^2$, namely rotations. When the center of rotation is $(0, 0)$, these again turn out to be linear transformations of $\mathbf{R}^2$.

Exercises

2.13. Let ρ_θ denote the counterclockwise rotation of $\mathbf{R}^2$ about $(0, 0)$ through the angle θ. Denote a typical point of $\mathbf{R}^2$ in polar coordinates as $(x, y) = (r\cos\alpha, r\sin\alpha)$.
  (a) What are the coordinates of the rotated point $\rho_\theta(x, y)$ in terms of $r, \theta,$ and α?
  (b) Write (x, y) as the column vector $u = \begin{bmatrix} x \\ y \end{bmatrix}$. Verify by elementary trigonometry that $\rho_\theta(x, y) = R_\theta u$, where
$$R_\theta = \begin{bmatrix} \cos\theta & -\sin\theta \\ \sin\theta & \cos\theta \end{bmatrix} \quad ∎$$

Although it is geometrically evident, we should check that reflections and rotations are in fact isometries of $\mathbf{R}^2$. This can of course be checked by "brute force," but a few observations and matrix notation will simplify the calculation a bit.

First recall the definition of the dot product of two vectors in $\mathbf{R}^2$:

If $u = (x, y)$ and $v = (x', y')$, then $u \cdot v = xx' + yy'$.

In particular, $u \cdot u = x^2 + y^2$, which is the square of the distance from u to $(0, 0)$. Since every positive real number has a unique square root, it is clear that a mapping $T : \mathbf{R}^2 \to \mathbf{R}^2$ is an isometry if and only if it preserves d^2, where d is the Euclidean distance function. But now

$$d^2(u, v) = (u - v) \cdot (u - v)$$

and

$$d^2(T(u), T(v)) = (T(u) - T(v)) \cdot (T(u) - T(v)).$$

If T is a linear transformation, we can rewrite this last equation as

$$d^2(T(u), T(v)) = T(u - v) \cdot T(u - v)$$

and so, setting $w = u - v$ we arrive at the fact:

Theorem. *A linear transformation $T : \mathbf{R}^2 \to \mathbf{R}^2$ is an isometry if and only if*

$$u \cdot u = T(u) \cdot T(u)$$

for all vectors $u \in \mathbf{R}^2$.

Now let A be the matrix representing the linear transformation T and let A^t be the transposed matrix; i.e., if

$$A = \begin{bmatrix} a & b \\ c & d \end{bmatrix},$$

then

$$A^t = \begin{bmatrix} a & c \\ b & d \end{bmatrix}.$$

Likewise if $u = \begin{bmatrix} x \\ y \end{bmatrix}$, then $u^t = (x, y)$, thought of as a 1×2 matrix. Now it is easy to see that

$$u \cdot u = u^t u,$$

where the second multiplication is matrix multiplication. Then it follows that
$$T(u) \cdot T(u) = (Au) \cdot (Au) = (Au)^t Au = u^t A^t Au.$$
Thus

Theorem. *Let $T : \mathbf{R}^2 \to \mathbf{R}^2$ be a linear transformation with associated matrix A. Then T is an isometry if $A^t A = I$, where*
$$I = \begin{bmatrix} 1 & 0 \\ 0 & 1 \end{bmatrix}.$$

Definition. We say that a square matrix A is an orthogonal matrix if $A^t A = I$.

Exercises

2.14. (Bonus) Prove that "if" may be replaced by "if and only if" in the preceding theorem.

2.15. Prove: The set $O(2)$ of all 2×2 orthogonal matrices with real entries is a group under matrix multiplication. (You may use the facts that matrix multiplication is associative and that $(AB)^t = B^t A^t$. Of course this formula is not so hard to check for 2×2 matrices. Why don't you check it?) This group is called the (two-dimensional) **orthogonal group**.

2.16a. Using the preceding theorem, verify that if r_ℓ is a reflection across a line ℓ passing through $(0, 0)$, then r_ℓ is an isometry of $\mathbf{R}^2$.

2.16b. Also verify that if ρ_θ is a rotation about $(0, 0)$, then ρ_θ is an isometry of $\mathbf{R}^2$. ∎

We would now like to show that every isometry R of $\mathbf{R}^2$ with $R(0, 0) = (0, 0)$ is either a reflection or a rotation about $(0, 0)$.

First we show

Lemma. *If R is any isometry of $\mathbf{R}^2$ with $R(0, 0) = (0, 0)$, then there is a reflection or a rotation R' such that*
$$R(0, 0) = R'(0, 0), \; R(1, 0) = R'(1, 0), \text{ and } R(0, 1) = R'(0, 1).$$

Proof. First, since R is an isometry fixing $(0, 0)$, $R(1, 0)$ and $R(0, 1)$ must lie on the unit circle. So we may set $R(1, 0) = (\cos \theta, \sin \theta)$ for some angle θ.

Next, using 2.1, we see that $R(0, 1)$ must lie both on the unit circle and on the line through $(0, 0)$ perpendicular to the vector $(\cos \theta, \sin \theta)$. Visually (or by an easy exercise) we know that there are exactly two such points: $(-\sin \theta, \cos \theta)$ and $(\sin \theta, -\cos \theta)$.

Clearly in the former case R maps both $(1, 0)$ and $(0, 1)$ in the same way as does the counterclockwise rotation through the angle θ about the point $(0, 0)$. In the latter case R maps both $(1, 0)$ and $(0, 1)$ in the same way as does the reflection across the line whose equation is
$$y = \tan\left(\frac{\theta}{2}\right) x.$$

This proves our claim. ∎

Now we have two isometries R and R', which agree on the three points $(0, 0)$, $(1, 0)$ and $(0, 1)$. Thus the isometry $S = R^{-1} \circ R'$ fixes these three points. What could S be?

Exercises

2.17. Prove: The set of points in $\mathbf{R}^2$ equidistant from two distinct points (a, b) and (c, d) is a straight line.

2.18. Prove: If S is an isometry of $\mathbf{R}^2$ fixing three noncollinear points A, B, and C, then S is the identity map. (*Hint*: If not, there is some point (a, b) such that $S(a, b) = (c, d)$ is distinct from (a, b). Now apply 2.17.) ∎

From the preceding exercises we immediately have the following theorem.

O(2) Theorem. *Every isometry of $\mathbf{R}^2$ that fixes the point $(0, 0)$ is either a rotation about $(0, 0)$ or is a reflection across a line through $(0, 0)$.*

Proof. By the preceding lemma, if R is any such isometry, then there is such a rotation or reflection R' that agrees with R on the three points $(0, 0)$, $(1, 0)$ and $(0, 1)$. Then $R^{-1} \circ R'$ fixes these three points. But then by 2.18

$$R^{-1} \circ R' = I$$

and so $R' = R$, as claimed. ∎

Combining the $O(2)$ Theorem with the Factorization Theorem, we get a general description of isometries of $\mathbf{R}^2$.

Isometry Theorem. *Let $S : \mathbf{R}^2 \to \mathbf{R}^2$ be an isometry with $S(0, 0) = (a, b)$. Then there is a rotation or reflection R such that $S = T_{(a,b)} \circ R$. Equivalently*

$$S(x, y) = R(x, y) + (a, b) \text{ for all } (x, y) \in \mathbf{R}^2.$$

In matrix/vector notation, if $u = \begin{bmatrix} x \\ y \end{bmatrix}$, then there is a 2×2 orthogonal matrix A such that

$$S(u) = Au + \begin{bmatrix} a \\ b \end{bmatrix}$$

We may attempt to extend this Isometry Theorem to higher dimensional spaces. The Factorization Theorem extends painlessly, but the structure of $O(n)$ becomes increasingly difficult to analyze as n increases. In particular not every isometry in $O(n)$ is a rotation or a reflection when $n \geq 3$. Euler gave a complete analysis for $n = 3$ without the benefit of linear algebra. One of the main uses of linear algebra has been to extend this study to higher dimensions without the prodigious computational prowess of Euler. We shall return to the study of $O(3)$ in Chapter 9.

There is yet another way to state the Isometry Theorem for $\mathbf{R}^2$. Let's represent $\mathbf{R}^2$ as the plane

$$\{(x, y, 1) : x, y \in \mathbf{R}\}$$

in $\mathbf{R}^3$. Then each Euclidean motion can be represented as a 3×3 matrix.
For instance if R is a rotation, and so the associated matrix is

$$A = \begin{bmatrix} \cos\theta & -\sin\theta \\ \sin\theta & \cos\theta \end{bmatrix},$$

then the entire Euclidean motion $S(u) = Au + \begin{bmatrix} a \\ b \end{bmatrix}$ is represented by the matrix

$$A^* = \begin{bmatrix} \cos\theta & -\sin\theta & a \\ \sin\theta & \cos\theta & b \\ 0 & 0 & 1 \end{bmatrix}$$

in the sense that if $u = \begin{bmatrix} x \\ y \end{bmatrix}$, then $S(u) = A^* \begin{bmatrix} x \\ y \\ 1 \end{bmatrix}$.

We will practice these ideas in the exercises, but first we recall the following concepts and facts from linear algebra.

Definition. Let $T : \mathbf{R}^n \to \mathbf{R}^n$ be a linear transformation. A nonzero vector v in $\mathbf{R}^n$ is an **eigenvector** for T if there is a scalar λ such that

$$T(v) = \lambda v.$$

We call λ the **eigenvalue** for T associated to the eigenvector v. Geometrically this means that the line passing through $\mathbf{0}$ and v is mapped into itself by T. If $\lambda > 0$, then the line is stretched (if $\lambda \geq 1$) or shrunk (if $\lambda < 1$) by the scaling factor λ. If $\lambda = 0$, then the line is collapsed to the point $\mathbf{0}$. If $\lambda < 0$, then the line is rotated $180°$ and then stretched or shrunk by the scaling factor $|\lambda|$.

If A is the matrix associated with the linear transformation T and if λ is an eigenvalue for T, then there is a nonzero eigenvector v such that

$$(A - \lambda I)v = 0.$$

Hence $A - \lambda I$ is a singular matrix and so $det(A - \lambda I) = 0$. Thus λ is a real eigenvalue for T if and only if λ is a real root of the polynomial $det(A - xI)$. This is the so-called **characteristic polynomial** of T.

From an earlier theorem we know that rotations about $(0, 0)$ are characterized as the unique isometries that fix $(0, 0)$ and no other point. There is nothing special about $(0, 0)$. In fact rotations are characterized as isometries of $\mathbf{R}^2$ that fix a unique point. Similarly reflections are characterized as the nonidentity isometries of $\mathbf{R}^2$ that fix two distinct points. We now apply this together with some linear algebra to help analyze the composite isometries of $\mathbf{R}^2$.

Exercises

2.18. Write the 3 × 3 matrix for the following motion: Rotate $\mathbf{R}^2$ 90° counterclockwise about (0, 0) and then translate by the vector $w = (1, 2)$. Find all real eigenvalues and eigenvectors for this matrix. Interpret this geometrically.

2.19. Using eigenvalue analysis, prove that any Euclidean motion of $\mathbf{R}^2$ of the form $T \circ \rho$ where T is a translation and ρ is a nonidentity rotation about (0, 0) is in fact a rotation.

2.20a. Prove: If u is a vector in $\mathbf{R}^2$, T_u is the translation by u, and ρ is any rotation about (0, 0), then

$$\rho \circ T_u = T_{\rho(u)} \circ \rho.$$

(You don't need matrices for this. Just recall that ρ is a linear transformation and recall the definition of T_u.)

2.20b. Using 2.19 and 2.20a, argue that any Euclidean motion of $\mathbf{R}^2$ of the form $\rho \circ T$, where T is a translation and ρ is a nonidentity rotation about (0, 0), is in fact a rotation.

2.21. Do the same type of eigenvalue/eigenvector analysis as in 2.18 for the following motion: Reflect $\mathbf{R}^2$ across the y-axis and then translate by the vector $v = (1, 0)$.

2.22. Do the same for the following motion: Reflect $\mathbf{R}^2$ across the y-axis and then translate by the vector $u = (0, 1)$.

2.23. Do the same for the following motion: Reflect $\mathbf{R}^2$ across the line $y = x$ and then translate by the vector $v = (1, 0)$.

2.24. Formulate a criterion for when a Euclidean motion of $\mathbf{R}^2$ of the form $T \circ R$ where T is a translation and R is a reflection is in fact a reflection. *Caution*: Not every isometry of the form $T \circ R$ where T is a translation and R is a nonidentity reflection fixes a point. The ones that do not fix *any* point are called **glide reflections**. They are *not* reflections.

For example, there is a symmetry of the frieze pattern illustrated here that is a glide reflection but not a reflection.

Bonus Exercise

2.25. Give a proof by picture of 2.19.

As remarked before this is not really a geometry book. But let's quickly finish up our project to formalize Euclid's SAS Theorem.

SAS Theorem. *Let ABC and DEF be two triangles with sides AB and DE of the same length, sides BC and EF of the same length, and angles ABC and DEF*

of the same measure. Then triangles ABC and DEF are congruent; i.e., there exists a Euclidean isometry of $\mathbf{R}^2$ *mapping A to D, B to E, and C to F.*

Sketch of a proof. Given any two points there is a translation mapping the first to the second. Thus we can find a translation T with $T(B) = E$. Let $T(A) = H$. As T is an isometry, the segments EH and ED have the same length, so there is a rotation ρ about E with $\rho(T(A)) = D$. Let $\rho(T(C)) = K$. As isometries preserve angle measure, the segments EK and EF make the same angle with the segment ED and also have the same length. Thus they either coincide or are mirror images of each other across the line ED. In the former case, $\rho \circ T$ is the desired isometry. In the latter case, $R \circ \rho \circ T$ is the desired isometry, where R is the reflection across the line ED. ∎

More Bonus Exercises

2.26a. In the context of the SAS Theorem, describe an alternative way to map triangle ABC onto triangle DEF using only reflections.

2.26b. Using 2.26a, prove that every Euclidean isometry of $\mathbf{R}^2$ can be written as the composition of one, two, or three reflections. (There is an analogous statement in n dimensions for $n \geq 2$ called the Cartan–Dieudonné Theorem.)

2.26c. Explicitly describe how to write any rotation as a product of two reflections. Likewise describe explicitly how to write any translation as a product of two reflections. What distinguishes the reflecting mirrors in these two cases?

CHAPTER 3

SYMMETRY

What immortal hand or eye
Dare frame thy fearful symmetry?

—William Blake

If F is a figure in the plane, then of course F is congruent to itself and this congruence is accomplished by the identity motion $I(x, y) = (x, y)$. However, sometimes there are more isometries of $\mathbf{R}^2$ that map F to itself. When this is the case, we say informally that F is "symmetric." More formally we can measure how symmetric F is.

Definition. Let F be any figure (set of points) in the Euclidean plane $\mathbf{R}^2$. The set of all isometries of $\mathbf{R}^2$ that map F onto itself is called the **symmetry group** of F, $\Sigma(F)$.

Exercise

3.1. Justify the terminology; i.e., prove that the set $\Sigma(F)$ of all isometries of $\mathbf{R}^2$ that map F onto F is a subgroup of $E(2)$, the group of all isometries of $\mathbf{R}^2$. (The Associative Law is automatic, since the operation is composition of functions. It suffices to verify that conditions (G1), (G2), and (G3) hold for $\Sigma(F)$.) ∎

If the figure F lies within a bounded region of the plane, then it is pretty clear that F has a "center of gravity" P and this point P must be fixed by every symmetry of F. (I won't attempt to prove this.)

First let's consider a bounded figure F with center of gravity at $(0, 0)$. Then the symmetry group $\Sigma(F)$ is a subgroup of the orthogonal group $O(2)$ of all Euclidean motions (rotations and reflections) of $\mathbf{R}^2$ fixing $(0, 0)$.

$O(2)$ itself is the symmetry group of the unit circle S^1. If P_n is a regular n-sided polygon inscribed in S^1, then $(0, 0)$ is the center of gravity of P_n and so $\Sigma(P_n)$ is a subgroup of $O(2)$. Furthermore $\Sigma(P_n)$ must map vertices of P_n to vertices of P_n. Let V_n be the set of vertices of P_n, identified with the set of numbers $\{1, 2, \ldots, n\}$.

Definition. A one-to-one (and necessarily onto) function
$$f : \{1, 2, \ldots, n\} \to \{1, 2, \ldots, n\}$$
is called a **permutation** of $\{1, 2, \ldots, n\}$.

The set of all permutations of $\{1, 2, \ldots, n\}$ clearly forms a group under composition of functions. The total number of permutations of $\{1, 2, \ldots, n\}$ is
$$n! = n \cdot (n - 1) \cdot \cdots 2 \cdot 1.$$

Definition. The **symmetric group** S_n is the group of all permutations of the set $\{1, 2, \ldots, n\}$.

Thus $|S_n| = n!$.

Theorem. *Let P_n be a regular n-sided polygon inscribed in S^1 for some $n \geq 3$. Let V_n be the set of all vertices of P_n. Fix a numbering of the vertices and identify $V_n = \{1, 2, \ldots, n\}$. Let $\Phi : \Sigma(P_n) \to S_n$ be the function defined by $\Phi(f) = \tilde{f}$, where $\tilde{f}$ is the restriction of the isometry $f : \mathbf{R}^2 \to \mathbf{R}^2$ to the domain $V_n = \{1, 2, \ldots, n\}$. Then Φ is a one-to-one function satisfying*
$$\Phi(f \circ g) = \Phi(f) \circ \Phi(g)$$
for all f and g in $\Sigma(P_n)$.

Proof. First we argue that Φ is a one-to-one function. Suppose on the contrary that f and g are symmetries of P_n with $\Phi(f) = \Phi(g)$. Then $f(v) = g(v)$ for every vertex $v \in V_n$. But $f(0, 0) = g(0, 0)$ and so $f^{-1} \circ g$ fixes $(0, 0)$ and every vertex $v \in V_n$. Since $f^{-1} \circ g$ fixes $(0, 0)$, $f^{-1} \circ g$ is either a rotation about $(0, 0)$ or a reflection across a line through $(0, 0)$. If it is a nonidentity rotation, then it fixes *no* vertex in V_n, which is not the case. If $f^{-1} \circ g$ is a reflection, then its fixed points all lie on one line through $(0, 0)$, which intersects the circle S^1 in exactly two points, contrary to the fact that P_n has $n \geq 3$ vertices. Hence $f^{-1} \circ g$ is the identity map, whence $f = g$, as desired.

Next we prove that $\Phi(f \circ g) = \Phi(f) \circ \Phi(g)$. Since all of the functions $\Phi(f)$, $\Phi(g)$, and $\Phi(f \circ g)$ have the same domain and range V_n, it suffices to check that the functions agree at a typical point $v \in V_n$. But by definition
$$\Phi(f \circ g)(v) = (f \circ g)(v) = f(g(v))$$
and
$$(\Phi(f) \circ \Phi(g))(v) = \Phi(f)(\Phi(g)(v)) = \Phi(f)(g(v)) = f(g(v)).$$
Thus the two functions $\Phi(f \circ g)$ and $\Phi(f) \circ \Phi(g)$ agree at every point of their common domains and so they are equal, as claimed. ∎

Definition. For each $n \geq 3$, we let D_n denote the image $\Phi(\Sigma(P_n))$ of the symmetry group $\Sigma(P_n)$ in the symmetric group S_n. We call D_n the **dihedral group** of degree n. (As defined here, D_n depends on the choice of numbering for the vertices of P_n.)

Definition. Let $(G, \circ)$ and $(H, \cdot)$ be two groups. A function $\Phi : G \to H$ is called an **isomorphism of groups** if Φ is a one-to-one function from G onto H satisfying

$$\Phi(g \circ g') = \Phi(g) \cdot \Phi(g') \text{ for all } g, g' \in G.$$

We then say that G and H are **isomorphic** as groups.

We may now restate the previous theorem as follows.

Theorem. $\Sigma(P_n)$ and D_n are isomorphic as groups.

This is an excellent example of the type of equivalence relation discussed at the end of Chapter 0. In many senses, $\Sigma(P_n)$ and D_n are completely different. The objects in $\Sigma(P_n)$ are functions defined on the entire Euclidean plane $\mathbf{R}^2$. The objects in D_n are permutations of the finite set $\{1, 2, \ldots, n\}$. But insofar as their "groupiness" is concerned, it is impossible to tell them apart. Any property of $\Sigma(P_n)$ that relates only to its nature as a group (e.g., its cardinality as a set, its nonabelianness) must hold true for D_n as well.

Exercises

3.2. Let G be a group with identity element e and let H be a group with identity element f. Suppose that $\Phi : G \to H$ is an isomorphism of groups.

3.2a. Prove: $\Phi(e) = f$. (Recall Exercise 1.15.)

3.2b. Prove: For all $g \in G$, $\Phi(g^{-1}) = (\Phi(g))^{-1}$. (Recall Exercise 1.17.)

3.3. Suppose that $\Phi : G \to H$ is an isomorphism of groups. Let e be the identity element of G and f be the identity element of H. Suppose that $g \in G$ and n is a natural number such that $g^n = e$. (Here we write $g^n = g \cdot g \cdot \cdots$ (n times).)

3.3a. Prove by induction on m:

$$\Phi(g^m) = (\Phi(g))^m \text{ for all } m \geq 1.$$

3.3b. Prove: $\Phi(g)^n = f$.

Now we consider some specific examples.

3.4a. Consider the equilateral triangle $T(=P_3)$ inscribed in the unit circle with one vertex at $(1, 0)$. List all the symmetries of T. Specifically let ρ denote the $120°$ counterclockwise rotation about $(0, 0)$ and let R denote the reflection across the x-axis. Write each symmetry of T as either I (the identity map) or as a "power" ρ^i of ρ or as $R \circ \rho^i$ for some i. Now write the "multiplication table" for the group $\Sigma(T)$. This is like an ordinary multiplication table with the rows and columns labeled by the elements of the group $\Sigma(T)$. The entry in row σ and column τ is the symmetry $\sigma \circ \tau$. Since $\Sigma(T)$ is a group, every "product" is again one of the elements of $\Sigma(T)$. Thus every entry in your multiplication table should be either I or R or ρ^i ($i = 1$ or 2) or $R \circ \rho^i$ ($i = 1$ or 2). Hence your job is to figure out, for example, another name for the symmetry $\rho \circ R$ to place in row ρ and column R.

Some permutation notation. We shall write

$$f = \begin{bmatrix} a & b & c \\ b & c & a \end{bmatrix}$$

to denote the permutation f of the set $\{a, b, c\}$ defined by $f(a) = b$, $f(b) = c$, and $f(c) = a$. In Chapter 8, we will introduce a more concise and useful notation for permutations, called cycle notation, introduced by Augustin Louis Cauchy.

3.4b. Label the vertices of T so that $(1, 0) = 1$, $\rho(1, 0) = 2$, and $\rho^2(1, 0) = 3$. Then under the map Φ, if f is a symmetry of T, $\Phi(f)$ is a permutation of $\{1, 2, 3\}$. For example

$$\Phi(\rho) = \begin{bmatrix} 1 & 2 & 3 \\ 2 & 3 & 1 \end{bmatrix}$$

and

$$\Phi(R) = \begin{bmatrix} 1 & 2 & 3 \\ 1 & 3 & 2 \end{bmatrix}.$$

List the six functions $\Phi(f)$ for each $f \in \Sigma(T)$. Figure out the multiplication table for this set of six functions. (This is the group D_3.) Now take the multiplication table for $\Sigma(T)$ that you found in 3.4a and replace f by $\Phi(f)$ everywhere in the table. Verify that this is exactly the same table as the one you just computed for D_3. (This is another proof that $\Sigma(T)$ and D_3 are isomorphic groups.) Notice that D_3 is the entire symmetric group S_3. This is a fluke. It is not true for D_n for any $n \geq 4$.

3.5a. Now consider the square Q inscribed in the unit circle with one vertex at $(1, 0)$. Again list all the symmetries of Q using only the symbols I, ρ (for the smallest rotation), and R. The multiplication table is getting too big to write down. How many entries would it have? Express $\rho \circ R$ in the form $R \circ \rho^i$ for some i. Is multiplication in the group $\Sigma(Q)$ a commutative operation? Explain how to compute the product of any two elements of $\Sigma(Q)$ using the basic rule you have derived.

3.5b. Again label the vertices of Q with the numbers $1, 2, 3, 4$, labeling counterclockwise starting at $(1, 0)$. With this numbering, express $\Phi(\rho)$ and $\Phi(R)$ as permutations of $\{1, 2, 3, 4\}$. Express $\Phi(\rho) \circ \Phi(R)$ in the form $\Phi(R) \circ \Phi(\rho)^i$ for some i by direct calculation of permutations. Verify that this is the same rule you would have predicted from 3.5a. As before, $\Phi(\Sigma(Q))$ is the dihedral group D_4. How many permutations are there in the group D_4? How many permutations are there in the symmetric group S_4?

3.6. Verify that the subset $\{I, R, \rho^2, R \circ \rho^2\}$ of $\Sigma(Q)$ is actually a subgroup of $\Sigma(Q)$. List the corresponding permutations $\Phi(I)$, $\Phi(R)$, $\Phi(\rho^2)$ and $\Phi(R \circ \rho^2)$. This subgroup of S_4 is often called the Klein 4-group V_4. Write the multiplication table for V_4. Verify that V_4 is an abelian group. The subgroup $\{I, R, \rho^2, R \circ \rho^2\}$ may be thought of as the full symmetry group of a (nonsquare) rectangle. Explain why.

3.7. Find a subgroup C_4 of D_4 such that $|C_4| = 4$ but C_4 and V_4 are not isomorphic as groups. *Prove* that they are not isomorphic. (*Hint*: Use Exercise 3.3.) Verify that C_4 is also an abelian group.

3.8. Now consider the regular n-gon P_n inscribed in the unit circle with one vertex at $(1,0)$. Do the same as for the square; i.e., give an economical description of all the symmetries of P_n. Then express $\rho \circ R$ in the form $R \circ \rho^i$ for some i. Explain how this rule permits the computation of any product of two symmetries of P_n.

3.9. Prove: For $n \geq 3$, D_n is isomorphic to S_n if and only if $n = 3$. ∎

Every figure F with a finite number of distinguished "vertices" must have a finite group of symmetries permuting the vertices and fixing the center of gravity of these vertices. Without loss we may assume that the center of gravity is the point $(0, 0)$. Thus $\Sigma(F)$ is a finite subgroup of $O(2)$. We would like to determine what all of the finite subgroups of $O(2)$ are. First we need another definition and a few observations.

Definition. $SO(2)$ is the set of all rotations of $\mathbf{R}^2$ about the point $(0, 0)$.

Exercises

3.10. Prove: $SO(2)$ is a subgroup of $O(2)$.

3.11. Prove: $SO(2)$ is the set of all linear transformations in $O(2)$ of determinant 1.

3.12a. Prove: The product of any two reflections R and R' in $O(2)$ is a rotation in $SO(2)$. (Describe the angle of rotation in terms of the angle between the reflecting mirrors.) What is the relationship between the rotation $R \circ R'$ and the rotation $R' \circ R$?

3.12b. Prove: The product of a reflection in $O(2)$ and a rotation in $SO(2)$ is a reflection in $O(2)$. ∎

First we will determine the finite subgroups of $SO(2)$, but before doing that, we need to discuss a certain class of groups briefly.

Definition. A group H is called a **cyclic group** if there is some element $h \in H$ such that every element of H can be written as h^r for some integer r. Here we define

$$h^0 = e \text{ and } h^{-r} = (h^r)^{-1} \text{ for all } r \in \mathbf{N}.$$

We say that H is generated by the element h. (In general h is not unique.) If h is a generator of H, we define the **order** of the element h to be the cardinality $|H|$ of the set H. Thus if H is an infinite group, we say that h is of infinite order.

Remark. *If G is any group and g is any element of G, then the subset $\{g^n : n \in \mathbf{Z}\}$ is easily seen to be closed under multiplication and inversion and contains $e = g^0$. (See the proof of the next theorem.) Hence it is a cyclic subgroup of the group G and is usually denoted $\langle g \rangle$. In particular we define the **order** of g to be $|\langle g \rangle|$ for any $g \in G$.*

In two of the theorems that follow we shall use the following basic property of the natural numbers $\mathbf{N}$.

The Well-Ordering Principle. *Let S be any nonempty subset of $\mathbf{N}$. Then S contains a smallest member s; i.e., there exists $s \in S$ with $n - s \geq 0$ for all $n \in S$.*

Theorem. *Suppose that H is a finite cyclic group; i.e., H is a cyclic group and $|H| = n < \infty$. If h is a generator of H, then $H = \{h, h^2, \ldots, h^n = e\}$, where e is the identity element of H. Thus if $1 \leq i, j \leq n$, then $h^i = h^j$ if and only if $i = j$.*

Proof. Since H is finite, the list $e = h^0, h = h^1, h^2, \ldots$ must have redundancies; i.e., there exist $i < j$ with $h^i = h^j$. Thus $j - i \in Z$, where
$$Z = \{n \in \mathbf{N} : h^n = e\}.$$

Choose m to be the smallest positive integer in Z. (*Note*: m exists by the Well-Ordering Principle.) If $0 \leq i < j < m$ with $h^i = h^j$, then $j - i \in Z$ with $j - i < m$, contrary to the choice of m. Thus
$$h = h^1, h^2, \ldots, h^m = e$$
is an irredundant list of elements of H.

Let h^i and h^j be on this list. Then
$$h^i h^{m-i} = h^m = e$$
and so $(h^i)^{-1} = h^{m-i}$ is on the list. If $i + j \leq m$, then clearly $h^i h^j = h^{i+j}$ is on the list. If $i + j > m$, then $1 \leq i + j - m \leq n$ and
$$h^i h^j = h^{i+j} = h^{i+j-m} h^m = h^{i+j-m} e = h^{i+j-m}.$$

So again $h^i h^j$ is on the list. Thus the set
$$H_0 = \{h, h^2, \ldots, h^m = e\}$$
is closed under multiplication and under inverses, and hence contains h^r for every integer r. But then $H_0 = H$ and $m = |H| = n$, completing the proof. ∎

Remark. *You are probably familiar with the Division Algorithm for the integers (though we shall review it later). If so you will easily verify that if H is a finite cyclic group with $|H| = n$ and with generator h, then if k is any integer, we may write $k = qn + r$ with $q, r \in \mathbf{Z}$ and $0 \leq r < n$. Then*
$$h^k = h^r \in \{e = h^0, h = h^1, h^2, \ldots, h^{n-1}\}.$$

Exercises

3.13. Prove: If H is a cyclic group with generator h, then h^{-1} is also a generator for H. Moreover $h = h^{-1}$ if and only if $|H| \leq 2$. (Thus H has more than one generator whenever $|H| > 2$.)

3.14a. Prove: $(\mathbf{Z}, +)$ is a cyclic group with generator 1.

3.14b. Verify that 1 and -1 are the only generators for $\mathbf{Z}$. ∎

Theorem.

If G and H are finite cyclic groups with $|G| = |H|$, then G and H are isomorphic groups.

If G is an infinite cyclic group, then $G \cong (\mathbf{Z}, +)$.

Proof. First let G and H be finite cyclic groups with $|G| = |H| = n$. Let g be a generator for G and let h be a generator for H. Let $\Phi : G \to H$ be the function $\Phi(g^i) = h^i$ for $1 \leq i \leq n$. Then clearly Φ is a one-to-one function from G onto H, and it remains to prove that
$$\Phi(g^i g^j) = \Phi(g^i)\Phi(g^j) = h^i h^j.$$
We have two cases to consider. If $i + j \leq n$, then $g^i g^j = g^{i+j}$ and $h^i h^j = h^{i+j}$. Then
$$\Phi(g^i g^j) = \Phi(g^{i+j}) = h^{i+j} = h^i h^j,$$
as desired. On the other hand, if $i + j > n$, then $g^i g^j = g^{i+j-n}$ and $h^i h^j = h^{i+j-n}$, and so again
$$\Phi(g^i g^j) = \Phi(g^{i+j-n}) = h^{i+j-n} = h^i h^j,$$
completing the proof for the finite case.

Next let G be an infinite cyclic group with generator g and let $\Phi : \mathbf{Z} \to G$ be the function $\Phi(n) = g^n$. Then clearly Φ is a surjective function and
$$\Phi(m+n) = g^{m+n} = g^m g^n = \Phi(m)\Phi(n).$$
It remains to argue that Φ is a one-to-one map. If $\Phi(m) = \Phi(n)$ with $m < n$, then
$$\Phi(m-n) = \Phi(m)\Phi(-n) = \Phi(m)\Phi(n)^{-1} = e.$$
Set $n - m = k > 0$. Then as in the previous proof, we may verify that the set $G_0 = \{g, g^2, \ldots, g^k = e\}$ is closed under inverses and under multiplication. But then since $g \in G_0$, $g^n \in G_0$ for every integer n and so $G = G_0$ is a finite set, contrary to assumption. Hence Φ must be a one-to-one map and so $\Phi : \mathbf{Z} \to G$ is an isomorphism of groups. ∎

Exercises

3.15a. Prove: $SO(2)$ contains finite cyclic subgroups of every possible cardinality n. Also $SO(2)$ contains infinite cyclic subgroups.

3.15b. (Bonus) Prove: $SO(2)$ is not a cyclic group. ∎

For our next theorem we will need to use both the Well-Ordering Principle for $\mathbf{N}$ and the following property of the real numbers, first explicitly used by Archimedes and hence often called the Archimedean Property of the Real Numbers:

If $\alpha \in \mathbf{R}$ with $\alpha > 0$, then there exists $n \in \mathbf{N}$ with $n\alpha \geq 1$.

(In other words there are no infinitesimally small real numbers.)

Theorem. *Every finite subgroup H of $SO(2)$ is a finite cyclic group of rotations. (Indeed if $|H| = n \geq 3$, then H is the group of all rotational symmetries of some regular n-gon.)*

Proof. Let H be a finite subgroup of $SO(2)$. If $H = \{I\}$, then obviously H is a cyclic group of rotations generated by I. Otherwise, we may consider each nonidentity rotation in H to be a counterclockwise rotation through some angle strictly between $0°$ and $360°$. Since $|H|$ is finite, we may choose the nonidentity rotation ρ in H through the smallest positive angle θ. We want to show that every rotation in H is a power of ρ, i.e., that every rotation in H is a rotation through an angle that is an integer multiple of θ.

Let σ be any nonidentity rotation in H and let ω be the angle of rotation. Then by the Archimedean Property of the Reals, there is some positive integer n such that $n\theta \geq \omega$. Let m be the smallest positive integer such that $m\theta \geq \omega$. (Again m exists by the Well-Ordering Principle.) Thus

$$(m-1)\theta < \omega \leq m\theta.$$

Let $\tau = \sigma^{-1} \cdot \rho^m$. Then $\tau \in H$ since H is a group. Also τ is the counterclockwise rotation through the angle $m\theta - \omega$. Now

$$0 \leq m\theta - \omega < m\theta - (m-1)\theta = \theta.$$

Since τ is a rotation in H through a nonnegative angle smaller than θ, we conclude that τ is the identity rotation I. Thus

$$\tau = \sigma^{-1} \circ \rho^m = I$$

and so $\sigma = \rho^m$.

Thus every rotation in H is a power of ρ, as claimed. Moreover we may apply the preceding argument to the identity rotation thought of as a rotation through the angle 2π to conclude that 2π is a multiple of θ. If $2\pi = n\theta$, then

$$H = \{\rho, \rho^2, \ldots, \rho^{n-1}, \rho^n = I\}.$$

Suppose $n \geq 3$. Let $p = (1, 0)$ and let V be the set of points

$$V = \{p, \rho(p), \rho^2(p), \ldots, \rho^{n-1}(p)\}.$$

Since ρ is an isometry,

$$d(p, \rho(p)) = d(\rho^i(p), \rho^{i+1}(p))$$

for all i. Thus V is the vertex set of a regular n-gon P_n inscribed in the unit circle of $\mathbf{R}^2$ and H is the group of all rotational symmetries of P_n. ∎

We now turn to the other finite subgroups of $O(2)$. One more exercise will be useful.

Exercises

Let G be a group with subgroups H and K.

3.16a. Prove: $H \cap K$ is a subgroup of G.

3.16b. Prove: $H \cup K$ is *not* a subgroup of G unless $H \leq K$ or $K \leq H$. ∎

Theorem. *Suppose that H is a finite subgroup of $O(2)$ that is not contained in $SO(2)$. Let $H_0 = H \cap SO(2)$ and let $|H_0| = n \geq 1$. Then one of the following conclusions holds:*

(a) $n = 1$ and $H = \{I, R\}$ *for some reflection* $R \in O(2)$; *or*

(b) $n = 2$ and H *is isomorphic to the Klein 4-group* V_4; *or*

(c) $n \geq 3$ and H *is the symmetry group of some regular n-gon and hence is isomorphic to the dihedral group* D_n.

Proof. Since H is not contained in $SO(2)$, H contains a reflection R. If ρ is any nonidentity rotation in H, then $R \circ \rho$ is a reflection in H different from R. Thus if R is the only reflection in H, then $H = \{I, R\}$ and we are done.

Suppose R' is another reflection in H. Then $\rho = R \circ R'$ is a rotation in H, and $\rho \neq I$, since $R^{-1} = R \neq R'$. Let H_0 be the set of all rotations in H. Then $H_0 = H \cap SO(2)$ and so H_0 is a finite subgroup of $SO(2)$ by the preceding exercise and so, by the previous theorem, H_0 is a finite cyclic group of rotations generated by the smallest nonidentity rotation in H_0, ρ. Then every rotation in H has the form ρ^i and every reflection R' in H satisfies $R \circ R' = \rho^i$ for some i. So every reflection in H has the form $R' = R \circ \rho^i$ for some i.

Moreover as $(R')^2 = I$ for every reflection R', we have
$$I = (R \circ \rho^i)^2 = R \circ \rho^i \circ R \circ \rho^i,$$
and so, using the fact that $R = R^{-1}$, we obtain
$$R \circ \rho^i = \rho^{-i} \circ R$$
for all i.

If $\rho^2 = I$, then $H = \{I, \rho, R, R \circ \rho\}$ and it is easy to verify that H is isomorphic to the Klein 4-group V_4. Finally assume that ρ has order $n \geq 3$. Let p be one of the points of intersection of the axis of symmetry of the reflection R and the unit circle. As before, let V be the set of points
$$V = \{p, \rho(p), \ldots, \rho^{n-1}(p)\}.$$
Then V is the vertex set of a regular n-gon P_n and $\rho \in \Sigma(P_n)$. We claim that also $R \in \Sigma(P_n)$. Indeed
$$R(\rho^i(p)) = \rho^{-i}(R(p)) = \rho^{n-i}(p) \in V,$$
proving the claim. ∎

Thus $H \leq \Sigma(P_n)$ and as $|H| = 2n = |\Sigma(P_n)|$, they must be equal as claimed.

As nothing is special about $(0, 0)$, we conclude that every finite planar symmetry group of cardinality unequal to 1, 2, or 4 is either the full symmetry group or the rotational symmetry group of a regular n-gon for some positive integer $n \geq 3$.

Exercise

3.17. Prove: If H is a planar symmetry group of cardinality 4, then either H is the full symmetry group of a (nonsquare) rectangle or H is the cyclic group of rotational symmetries of a square. ∎

A symmetry group of cardinality 2 is either $\{I, R\}$ for some reflection R or $\{I, \rho\}$ where ρ is a 180° rotation. For completeness and uniformity we write $D_2 = V_4$, $D_1 = \{I, R\}$, $C_2 = \{I, \rho\}$, and $C_1 = \{I\}$. Note that D_1 and C_2 are isomorphic as abstract groups. This is the only abstract isomorphism among the groups C_n and D_n.

Thus we have two infinite classes of finite symmetry groups: the finite cyclic groups of rotational symmetries
$$C_1, C_2, C_3, \ldots$$
and the finite dihedral groups of rotational and reflectional symmetries
$$D_1, D_2, D_3, \ldots.$$

We shall see that these provide examples of all of the finite groups of cardinality less than 8. Indeed these provide examples of all of the finite groups of cardinality p or $2p$ for p a prime number.

Two-dimensional symmetry groups are of interest in art and architecture. As such they were studied (in some sense) by Leonardo da Vinci. Hermann Weyl writes (using the terminology "improper rotations" for reflections):

> The first pure central building after antiquity, S. Maria degli Angeli in Florence (begun 1434) is an octagon. ... Leonardo da Vinci engaged in systematically determining the possible symmetries of a central building and how to attach chapels and niches without destroying the symmetry of the nucleus. In abstract modern terminology, his result is essentially our above table of the possible finite groups of rotations (proper and improper) in two dimensions.

Before leaving this chapter on symmetry, it behooves us to mention the most famous of all symmetrical objects: the Platonic solids.

Definition. A **Platonic solid** or **regular polyhedron** is a closed bounded solid figure whose boundary is a union of congruent regular polygons, meeting pairwise at an edge, and such that the same number of polygonal faces meet at each vertex.

The remarkable fact is that although there are infinitely many regular polygons, the regular polyhedra are very limited. This is not too hard to see, if we make two elementary observations:

1. Since two polygonal faces can meet in at most one edge, the number r of faces meeting at a vertex must satisfy $r \geq 3$.
2. The sum of the angles meeting at a vertex must be less than $360°$, since the figure must be convex at each vertex.

Exercise

3.18. Find a formula for the angle measure (in degrees) of any angle of a regular n-gon (n-sided polygon). Use this to determine the five possible configurations of polygons meeting at a vertex of a regular polyhedron. ∎

Now that you have determined the five possible local configurations, there are two questions:

(a) *Does each local configuration determine at most one regular polyhedron?*

(b) *Does each local configuration determine at least one regular polyhedron?*

Question (a) is essentially answered by the following global formula, due to Descartes:

Theorem. *If P is any closed connected figure in 3-space bounded by polygonal faces and having V vertices, E edges, and F faces, then*

$$V - E + F = 2.$$

By doing a Mercator-style projection into the plane, we see that this theorem is equivalent to the following theorem.

Theorem. *If G is any connected planar graph having V vertices, E edges and F faces, then if the outside is counted as a face, we have*

$$V - E + F = 2.$$

This theorem can be established inductively. It is clearly true for the graph with one vertex and no edges. Given any planar connected graph, removing an edge does one of two things:

(a) *joins two faces, thereby reducing F by 1; or*
(b) *disconnects the graph into two connected graphs.*

In the first case, as V is unchanged and both E and F are decreased by 1, we see that $V - E + F$ is unchanged. In the second case, let the numbers for the two connected subgraphs be V_1, E_1, F_1 and V_2, E_2, F_2. Then $V = V_1 + V_2$, $E = E_1 + E_2 + 1$ (since an edge was deleted), and $F = F_1 + F_2 - 1$ (since the outside was counted twice). By induction

$$V_1 - E_1 + F_1 = 2$$

and

$$V_2 - E_2 + F_2 = 2.$$

Hence

$$(V_1 + V_2) - (E_1 + E_2) + (F_1 + F_2) = 4$$

and so

$$V - E + F = 2,$$

as claimed.

Thus the formula is valid by induction on the number E of edges. ∎

Exercises

3.19a. Let P be a polyhedron in which r edges meet at each vertex. Let V be the total number of vertices and E the total number of edges. Imagine a table whose rows are labeled by the vertices and whose columns are labeled by the edges. Place a 1 in the (v, e) position of the table if vertex v lies on edge e. Notice that each row has r 1's and each column has two 1's. Use this to get a formula relating V, E, and r.

3.19b. Now let P be a polyhedron in which each face has exactly n edges. Let F be the total number of faces and E the total number of edges. Get a formula relating F, E, and n.

3.19c. Use Descartes' Formula along with 3.19a and 3.19b to show that for each "local" configuration of a possible regular polyhedron, the total number of vertices, edges, and faces is uniquely determined. Find these numbers.

Exercise 3.19 pretty much shows that there are at most five regular polyhedra. Book XIII of Euclid's *Elements* is devoted to explicit constructions showing that each of the five Platonic solids does in fact exist. We shall revisit this theme in Chapter 9, when we study the finite groups of rotations in $\mathbf{R}^3$.

Again as noted before, the hidden motive of these two chapters was to explore the concept of a group. Möbius in 1827 was studying groups of motions of the plane at the same time as Cauchy and Galois were studying groups of permutations of the roots of an equation, as we shall soon see. It took several decades before the mathematical community realized that these were two manifestations of the same mathematical beast. Permutations of a finite set don't look a lot like functions on an infinite space of points. It is only by "squinting" so that you only see their "groupiness" that you realize these are very much the same things. When this realization became "commonplace," shortly after the middle of the 19th century, mathematics took a great leap forward.

Let's briefly return to the title question: What is a geometry?

One way to think about the fundamentals of geometry is to say that a geometry Γ is a space P of "points" with a distance function (or **metric**) $dist : P \times P \to \mathbf{R}^+$ assigning to each pair of points a and b in P a nonnegative real number $dist(a, b)$, the distance between them. Of course $dist$ should have some "good" properties in order to be an honest distance function. Once we have this, we get a group of functions, $I(P)$, the group of all isometries of P. This group in turn gives us a definition of congruence of figures in our geometry Γ. Circles of course are the loci of all points equidistant from a given point a. Straight lines can be "recovered" as paths of minimal length (shortest distance, at least locally) between two points (geodesics).

Alternatively we may think of a geometry as a space of points on which a group acts "homogeneously" and use the group to define what we should mean by distance on the space. This was the point of view championed by Felix Klein in his celebrated Erlanger Programm in 1872, elaborating on work of Arthur Cayley.

We shall return occasionally to geometry, but for now, following the lead of the Arab mathematicians of the Middle Ages and their European successors, we shall respectfully leave geometry where the Greeks left it and turn our attention to algebra.

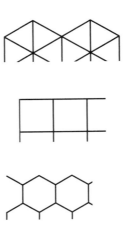

SECTION TWO
Polynomials

CHAPTER 4   THE ROOT OF IT ALL

CHAPTER 5   THE RENAISSANCE OF ALGEBRA

CHAPTER 6   COMPLEX NUMBERS

CHAPTER 7   SYMMETRIC POLYNOMIALS AND THE FUNDAMENTAL THEOREM OF ALGEBRA

CHAPTER 8   PERMUTATIONS AND LAGRANGE'S THEOREM

CHAPTER 9   ORBITS AND CAUCHY'S FORMULA

CHAPTER

4

THE ROOT OF IT ALL

Dating the beginning of mathematics even to the nearest millenium is impossible without some agreement as to the nature of mathematics. Certainly the origins of counting are lost in the mists of prehistory. Even the beginnings of "experimental" geometry, algebra, and trigonometry are very remote. However what most mathematicians regard as mathematics is not solely the "empirical" discovery of certain truths but the rigorous verification of these truths by the means of a certain formalized deductive reasoning called "mathematical proof" for which Euclid's *Elements* served for centuries as a paradigm.

The psychological and sociological factors that precipitated the development of mathematics in this sense in the Greek world beginning around 500 B.C. are beyond our competence to discuss. Suffice to say that surely one impetus was the discovery of a fact or facts of striking beauty and simplicity, discovered empirically and yet so unexpected as to demand "proof." Almost certainly one such fact is the celebrated **Pythagorean Theorem**:

> Given a right triangle with legs of lengths a and b and hypotenuse of length c, we have $a^2 + b^2 = c^2$.

At a computational level this theorem was known to the Babylonians and to the Chinese centuries before Pythagoras and may have been one motivation for the investigation of solutions of quadratic equations. In any case an algorithm for solving quadratic equations was known to the Babylonians, probably in the following form:

Suppose we are trying to find two numbers h and w whose sum $b = h + w$ and whose product $c = hw$ are known. (If this seems too artificial, you might prefer the formulation: Suppose we are trying to find the dimensions h (height) and w (width) of a field whose perimeter $2b$ and area c are known.)

If we can find $d = p - q$, then by averaging b and d we can find h, and then w. But

$$d^2 = (p-q)^2 = p^2 - 2pq + q^2 = (p^2 + 2pq + q^2) - 4pq$$
$$= (p+q)^2 - 4pq = b^2 - 4c.$$

So we have reduced the general quadratic equation problem
$$x^2 - bx + c = 0$$
to the special problem of taking a square root, i.e., solving for d in the equation
$$d^2 = b^2 - 4c.$$

If $b^2 - 4c$ is negative, the traditional attitude was that there were no solutions. Later people would come to terms with these "imaginary" solutions.

Exercises

4.1a. If 5 is one root of the quadratic equation $x^2 + 7x + c = 0$, what is the other root? (Think about this. You should be able to answer this without knowing c.) What is c?

4.1b. In general if r is one root of the quadratic equation $x^2 + bx + c = 0$, what is the other root s (in terms of r and b)? What is c (in terms of r and s)?

Centuries later Diophantus (circa 300 A.D.) was interested in finding rational solutions of equations. Specifically he looked at the Pythagorean Equation:
$$a^2 + b^2 = c^2.$$

Geometrically this is related to the problem of finding points (x, y) on the unit circle S^1 with both x and y rational numbers.

4.2. Justify the following assertions: Any triple (a, b, c) of rational numbers satisfying the Pythagorean Equation other than the triple $(0, 0, 0)$ corresponds to a rational point $(a/c, b/c)$ on the unit circle S^1. Conversely any rational point (x, y) on the unit circle S^1 corresponds to infinitely many triples (a, b, c) of rational numbers satisfying the Pythagorean Equation. In fact (x, y) will correspond to infinitely many triples (a, b, c) of positive integers satisfying the Pythagorean Equation.

4.3a. Prove: If (x, y) is a rational point on the unit circle other than $(-1, 0)$ and L is the line (chord) passing through $(-1, 0)$ and (x, y), then L has rational slope. (What is the slope?)

4.3b. Conversely prove: If L is any line through $(-1, 0)$ with rational slope t, then L intersects S^1 in a second rational point. (*Hint*: Write the equation for L in "point-slope" form using the slope t. Plug in for y in the equation for S^1 to get a quadratic equation in x. You *know* one root of this equation. Use the method of Exercise 4.1 to find the second root (in terms of t).)

4.4. From 4.3, you should have formulas for $x = a/c$ and $y = b/c$ in terms of t, where (x, y) is a rational point on the circle S^1. Let $t = s/r$ where r and s are positive integers. Plug in and derive the formula for an integer Pythagorean triple:
$$a = r^2 - s^2, b = 2rs, c = r^2 + s^2.$$
Verify that $a^2 + b^2 = c^2$. What additional condition on the positive integers r and s is needed to guarantee that a, b, and c are the sides of some right triangle?

4.5a. List the six smallest examples of integer Pythagorean triples.

4.5b. Which of the integers $\{1, 2, \ldots, 13\}$ can be the length of the hypotenuse of a right triangle whose legs have integer lengths? Justify your answer.

4.5c. Formulate a conjecture concerning which positive integers can be the length of the hypotenuse of a right triangle whose legs have integer lengths.

4.6. Using 4.3 and the fact that there are rational numbers arbitrarily close to any given real number, give a *plausible* argument (not a proof) for the following statement:

> Given any angle θ less than $90°$, there is a right triangle one of whose angles is very close in size to θ and all of whose sides have integer length. ∎

Remark. *The ancient Babylonians were aware of this fact and were able to produce vast quantities of Pythagorean Triples (triples of positive integers (a, b, c) such that $a^2 + b^2 = c^2$) and use them to construct "trig tables" to assist them in doing astronomy.*

CHAPTER 5

THE RENAISSANCE OF ALGEBRA

The methodology of what we would now call "high school algebra" though not with our current notation, was carefully developed and codified by the Hindu, Arab, and Persian mathematicians of the 6th to 12th centuries. One of the high points was the quadratic formula for the solution of second-degree equations, which was formulated in essentially its modern form.

Of course the key step in the solution is the removal of the first-degree term by the method of "completing the square":

Given the equation

$$x^2 + bx + c = 0,$$

we introduce the new variable $y = x + \frac{b}{2}$ and substitute for x to get the new equation

$$y^2 = \frac{b^2}{4} - c,$$

which can be solved for y by the extraction of square roots. We then can easily recover x.

Buoyed by our success with quadratic equations, we try to apply the same method to the cubic equation:

$$x^3 + ax^2 + bx + c = 0.$$

Historically examples were already considered by the early Hindu mathematicians, and a geometric method of solution (as the intersection point of a parabola and a hyperbola) was known to the Persian mathematician Omar Khayyam in the 12th century A.D.

Exercise

5.1. Find a change of variables that reduces this equation to the "easier" equation:

$$y^3 - 3py + q = 0,$$

where p and q are certain constants expressible in terms of the original coefficients a, b, and c. (*Hint*: Think about what made the change of variables in the quadratic equation work.) ∎

So far so good, but it is not obvious that the new equation is easier to solve than the old one. We can't eliminate the y term by a simple change of variables without reintroducing a quadratic term. Or can we?

Exercise

5.2. Try it. Use a similar trick to the one you used in 5.1 to eliminate the linear term from the equation
$$y^3 - 3py + q = 0.$$
What happens? ∎

A new idea was needed and finally was discovered (probably semi-independently) by some Italian mathematicians—Scipione del Ferro and Niccolo Fontana (known as Tartaglia)—in the early 16th century and published by Girolamo Cardano in a book entitled *Ars Magna*.

Let's go back to calling the variable x. We have reduced our cubic equation to the simpler equation
$$x^3 - 3px + q = 0,$$
or

(5A) $$x^3 - 3px = -q.$$

The new idea needed was probably inspired by the following expansion:
$$(u+v)^3 = u^3 + 3u^2v + 3uv^2 + v^3 = u^3 + 3uv(u+v) + v^3,$$
whence
$$(u+v)^3 - 3uv(u+v) = u^3 + v^3.$$
So if we set $x = u + v$, this becomes
$$x^3 - 3(uv)x = u^3 + v^3.$$
The problem now is to solve the pair of equations:
$$uv = p$$
(5B) $$u^3 + v^3 = -q$$
for u and v (after which we easily will find $x = u + v$). The obvious next move is to set $v = p/u$ from the first equation and substitute in the second, obtaining
$$u^3 + \frac{p^3}{u^3} = -q.$$
Multiplying by u^3, we end up with the new equation:
$$u^6 + qu^3 + p^3 = 0.$$

THE RENAISSANCE OF ALGEBRA   51

This seems to be really bad news. We started with a cubic equation and after all this work we now have an equation of degree 6. (Remember the number 6 (= 3!). This was one of the clues that showed Lagrange what was going on.)

We seem to be moving in the wrong direction. But then we notice that our new equation is actually a quadratic equation in u^3. So we can solve for u^3, then for u, and finally for x.

The alert reader may be concerned about whether we have really defined u at all.

Exercise

5.3. Solve the pair of equations

$$u + v = x$$
$$uv = p$$

for u in terms of p and x. (There are actually two possibilities.) Is u ever equal to 0? (This would be bad, since $x = u + (p/u)$.) Why is this not a problem? Assuming that x is always a real variable, is u always real? At what point does it "change"? ■

Rather than write down the rather horrifying formula (usually referred to as Cardano's Formula) that results, let's try to apply this in a specific example. Let's just start with a reduced equation:

$$x^3 + 3x - 4 = 0.$$

It suffices to remember the trick of setting $x = u + v$ and the goal of eliminating all terms except u^3, v^3 and the constant term. Then just follow your nose:

$$(u+v)^3 + 3(u+v) - 4 = u^3 + v^3 - 4 + (3uv + 3)(u+v) = 0.$$

Setting $uv = -1$, we have

$$u^3 + v^3 - 4 = 0,$$

and then, since $v = -1/u$,

$$u^3 - \frac{1}{u^3} - 4 = 0.$$

So

$$u^6 - 4u^3 - 1 = 0.$$

Then by the quadratic formula

$$u^3 = \frac{4 \pm \sqrt{16 + 4}}{2} = 2 \pm \sqrt{5}.$$

We end up with two horrible looking solutions:

$$x = \sqrt[3]{2 + \sqrt{5}} - \frac{1}{\sqrt[3]{2 + \sqrt{5}}},$$

and

$$x = \sqrt[3]{2-\sqrt{5}} - \frac{1}{\sqrt[3]{2-\sqrt{5}}}.$$

This is particularly disturbing because there is a solution to our cubic equation that is fairly obvious by inspection.

Exercises concerning $x^3 + 3x - 4$

5.4. Find the "obvious" solution of $x^3 + 3x - 4 = 0$.

5.5. Using calculus determine how many real solutions there are of $x^3 + 3x - 4 = 0$. What do you conclude about $\sqrt[3]{2+\sqrt{5}} - 1/\sqrt[3]{2+\sqrt{5}}$?

5.6. Using this, find an expression for $\sqrt[3]{2+\sqrt{5}}$ that does not involve cube roots.

5.7. Likewise find an expression for $\sqrt[3]{2-\sqrt{5}}$ that does not involve cube roots. ∎

It's not as easy to tell if a number is irrational as we might think!

Armed with one solution of $x^3 + 3x - 4 = 0$, we can easily find all of the solutions. Let's recall why.

Exercise

5.8. Prove by mathematical induction on n: If a is any number, then $x - a$ is a factor of the polynomial $x^n - a^n$. More specifically

$$x^n - a^n = (x-a)q_n(x),$$

where $q_n(x)$ is a polynomial of degree $n - 1$ with leading coefficient 1 and with all other coefficients expressible as polynomials in a with integer coefficients. (*Hint*: Do the first step of the polynomial division of $x^{n+1} - a^{n+1}$ by $x - a$. What is the remainder polynomial? Apply induction.) ∎

Definition. A **ring of numbers** R is a set of (real) numbers containing 0 and closed under the operations of addition, subtraction and multiplication. If R is also closed under division by nonzero members of R, then R is called a **field of numbers**.

Clearly if R is a ring of (real) numbers and a is a (real) number, then the set $R[a]$ of all numbers of the form

$$p(a) = c_0 + c_1 a + \cdots + c_n a^n \text{ with } c_i \in R \text{ and } n \in \mathbf{N}$$

is a ring of numbers, being again closed under addition, subtraction, and multiplication.

Factor Theorem. *Let $p(x)$ be a polynomial of degree n with coefficients in the ring R of numbers. Suppose that a is a root of $p(x)$; i.e., $p(a) = 0$. Then*

$$p(x) = (x-a)q(x)$$

for some polynomial $q(x)$ of degree $n - 1$ with coefficients in the ring $R[a]$.

Proof. Let
$$p(x) = c_0 + c_1 x + \cdots + c_n x^n.$$
Then since $p(a) = 0$,
$$p(x) = p(x) - p(a) = (c_0 - c_0) + c_1(x - a) + c_2(x^2 - a^2) + \cdots + c_n(x^n - a^n).$$
By 5.8,
$$x^m - a^m = (x - a)q_m(x),$$
where $q_m(x)$ is a polynomial of degree $m - 1$ with coefficients in $R[a]$. Thus
$$p(x) = (x - a)[c_1 + c_2 q_2(x) + \cdots + c_n q_n(x)] = (x - a)q(x),$$
where $q(x)$ is a polynomial of degree $n - 1$ with coefficients in $R[a]$, as claimed. ∎

The Factor Theorem appeared first in *La Géometrie* of René Descartes (1637). As a corollary we obtain the following interesting result, using the **Cancellation Property** of ordinary numbers:

Let a and b be (real) numbers. Then
$$ab = 0 \text{ if and only if } a = 0 \text{ or } b = 0.$$

Corollary to the Factor Theorem. *If $p(x)$ is a polynomial with real coefficients of degree $n \geq 1$, then $p(x) = 0$ has at most n real solutions.*

Proof. We proceed by mathematical induction on n. If $n = 1$, then $p(x) = ax + b$ has the unique solution $x = -b/a$. Now suppose that $n > 1$ and suppose that the theorem is true for polynomials of degree $n - 1$. If $p(x) = 0$ has no real solutions, then we are done. On the other hand, if $x = a$ is a real solution, then by the Factor Theorem
$$p(x) = (x - a)q(x)$$
for some polynomial $q(x)$ of degree $n - 1$ with real coefficients. Since $q(x)$ has degree $n - 1$, the inductive hypothesis tells us that $q(x) = 0$ has at most $n - 1$ real solutions.

Now suppose that b is a real solution of $p(x) = 0$ different from a. Then
$$0 = p(b) = (b - a)q(b).$$
Thus $b - a$ and $q(b)$ are two real numbers whose product is 0. By the Cancellation Law, since $b - a \neq 0$, we must conclude that $q(b) = 0$. Hence every solution of the equation $p(x) = 0$ with the possible exception of $x = a$ is a solution of $q(x) = 0$. Since $q(x) = 0$ has at most $n - 1$ real solutions by the inductive assumption, $p(x) = 0$ has at most n real solutions as claimed. ∎

Remark. *There is another fact about "ordinary numbers" that plays a subtle but fundamental role here, namely the Commutative Law for Multiplication. William Rowan Hamilton introduced a system of "numbers" where all of the ordinary algebra rules hold except for the Commutative Law for Multiplication. In the ring of quaternion numbers, the equation $x^2 + 1 = 0$ has infinitely many solutions. We shall study Hamilton's quaternions a bit in Chapter 9A.*

Of course a polynomial of degree n with real coefficients need not have n real roots. Indeed the same equation $x^2 + 1 = 0$ has *no* real solutions.

Exercise

5.9. Using the "obvious" root to get started, find all roots of $x^3 + 3x - 4$. ∎

Besides the fact that Cardano's Formulas give really ugly expressions for really nice roots of some cubics, an even bigger problem was recognized by Cardano and his contemporaries. Sometimes the formulas don't appear to give *any* solution in the real numbers even though a real solution is known to exist. This situation was referred to as the "**casus irreducibilis**." Let's look at an example:

$$x^3 - 6x + 5 = 0.$$

Now $p = 2$, $q = 5$, and our reduction method leads to the quadratic equation in u^3:

$$(u^3)^2 + 5u^3 + 8 = 0.$$

The discriminant of this quadratic is $q^2 - 4p^3 = 5^2 - 4 \cdot 8 = -7$, so the quadratic has no (real) solutions. However, by inspection the cubic *does* have real solutions.

Exercises

5.10. Find all solutions of $x^3 - 6x + 5 = 0$.

5.11. Find all solutions of $x^3 - 6x + 4 = 0$. Verify that this equation also falls in the "casus irreducibilis."

5.12. Give an argument (from calculus) that every cubic equation with real coefficients has at least one real solution. ∎

Remark. *Later we shall speak of certain polynomials as "irreducible polynomials." Although the Latin root is the same, this has no mathematical relationship to the term* casus irreducibilis.

This impasse finally forced mathematicians to come to terms with the complex numbers and their arithmetic. However, before that there was an interesting alternative trick discovered by Francois Viete. We outline it in the following exercises.

Exercise

Consider the cubic equation $x^3 - 3px + q = 0$.

5.13. Verify that the casus irreducibilis for this equation holds precisely when $q^2 < 4p^3$. In particular $p > 0$. Prove that in the casus irreducibilis, the equation *always* has three real roots and they lie between $-2\sqrt{p}$ and $+2\sqrt{p}$. (Use calculus.) ∎

Now we can describe Viete's Method for solving the equation

$$x^3 - 3px + q = 0$$

in the casus irreducibilis.

Recall the addition formula for sine and cosine:
$$\sin(A+B) = \sin(A)\cos(B) + \sin(B)\cos(A)$$
and
$$\cos(A+B) = \cos(A)\cos(B) - \sin(A)\sin(B).$$
When $A = B$ we get (using the Pythagorean Formula $(\cos^2(A) + \sin^2(A) = 1)$:
$$\cos(2A) = \cos^2(A) - \sin^2(A) = 2\cos^2(A) - 1$$
and
$$\sin(2A)\sin(A) = 2\sin(A)\cos(A)\sin(A) = 2\cos(A)\sin^2(A) = 2\cos(A) - 2\cos^3(A).$$
Proceeding further we get
$$\cos(3A) = \cos(2A + A) = \cos(2A)\cos(A) - \sin(2A)\sin(A) = 4\cos^3(A) - 3\cos(A).$$
This cubic equation in cos:
$$4\cos^3(A) - 3\cos(A) = \cos(3A)$$
was Viete's inspiration.

Since $p > 0$ in the casus irreducibilis, Viete defined a new variable w by
$$x = 2\sqrt{p}w.$$
This amounted to "rescaling" the polynomial $x^3 - 3px + q$ so that its roots now lie between -1 and $+1$. After substitution the new cubic equation is
$$4w^3 - 3w = c,$$
where $c = \frac{q}{2p\sqrt{p}}$. This gives a trigonometric solution of the equation:
$$x = 2\sqrt{p}\cos\left(\frac{\cos^{-1}\left(\frac{q}{2p\sqrt{p}}\right)}{3}\right).$$

Exercises

5.14. Verify that $|c| < 1$ precisely in the casus irreducibilis.

5.15. Using these ideas, solve the following cubic equations,
$$x^3 - 3x - 1 = 0$$
and
$$x^3 - 12x - 8 = 0,$$
in the sense that you

(a) Express the roots in terms of cosines of specific angles.

(b) Use your handy calculator (or trig table—do they still exist?) to get **approximate** numerical values for the roots.

(c) Draw a picture of the unit circle and illustrate the relevant angles, their cosines, and the roots of the cubic equation.

5.16. Discuss the following statement with respect to the specific equations in 5.15:

There are infinitely many angles with a given cosine. Why don't you get infinitely many different roots for the cubic equation? How do you know when to stop?

5.17. (Digression) In passing from $x^3 - 3px + q$ to $4w^3 - 3w - c$, we passed from a two-parameter family of polynomials to a one-parameter family. How is that possible? Did we lose something? Can this be used to prove that **R** and **R**^{2} have the same cardinality? ∎

More Exercises (Elliptic Curves)

One of the most important family of curves historically is the family of **elliptic curves**. These include all plane curves whose equation is of the form

$$y^2 + ay = bx^3 + cx^2 + dx + e, \quad b \neq 0.$$

These curves were studied extensively by Newton. A deeper study was undertaken by Gauss, Abel, and Jacobi. Most recently (and surprisingly) they played a crucial role in Wiles' proof of Fermat's Last Theorem. We consider only elliptic curves over **Q**, i.e., those with integer coefficients a, b, c, d, and e.
Let's look a little at one example: the curve E whose equation is

$$y^2 = x^3 - 3x - 1.$$

5.18. Find the points where this curve intersects the x-axis.

5.19. Draw some pictures to illustrate the family of cubic curves:

$$y = x^3 - 3x - (1 + c^2)$$

as c^2 increases. (Note that the left side of the equation is y not y^2. These are ordinary cubic equations, not the equations of elliptic curves.) What happens to the real roots of these polynomials as c^2 increases? At what "point" does the polynomial stop having three real roots? Use this and symmetry about the x-axis to sketch the elliptic curve E.

5.20. If -1 and 3 are two roots of the cubic equation $x^3 + 5x^2 + ex + f = 0$, what is the third root? What are e and f? (Again you should be able to answer the first question first.)

5.21. (Diophantus' Chord and Tangent Method) Suppose $P = (r, s)$ and $Q = (t, u)$ are two rational points on an elliptic curve E. If $P \neq Q$, prove that if the chord through these two points meets the elliptic curve in a third point R, then R is also a rational point. If $P = Q$, prove that if the tangent line to E at P meets E in a second point R, then R is a rational point. Explain how you can find the coordinates of R *without* solving a cubic equation.

Definition. Let P and Q be two rational points on the elliptic curve E. If the chord through P and Q (respectively, the tangent line to E at P) meets E in another point R, define $P + Q$

(respectively $2P$) to be the "mirror image" of R across the (horizontal) axis of symmetry of the elliptic curve E. (*Note*: $P + Q$ is also a rational point of E.)

5.22. Apply this to the elliptic curve $E' : y^2 + y = x^3 - x$ starting with the "obvious" rational points $P = (0, 0)$ and $Q = (1, 0)$. Specifically setting $S = P + Q$, find $S, 2S = (P + Q) + S, P + (Q + S)$, and $(S + P) + Q$. (It is one of the amazing facts of mathematics that this addition rule on any elliptic curve E satisfies the Associative Law (as well as the Commutative Law) and so defines an abelian group structure on the set $E(Q)$ of all rational points of C if you throw in the "point at infinity" ∞ as the identity element in the group. Guess what is the inverse of the point P.) ∎

Around 1540 Cardano's student, Lodovico Ferrari, found a method for reducing the problem of solving quartic (fourth-degree) equations to the problems of solving quadratic and cubic equations. His method was a clever "completing the square" trick outlined in the exercises that follow. This solution method was devised in response to a specific challenge problem posed by Zuanne da Coi:

Divide 10 into three parts a, b and c such that $ac = b^2$ and $ab = 6$.

At the time quartic (and higher degree) equations were not considered of serious interest because algebra was still tied intimately to geometry: "squares" (x^2) were areas, "cubes" (x^3) were volumes; since there was no higher geometric dimension imaginable, higher powers had no "meaning." Even the notation (x^2, x^3, etc.) did not yet exist. All this was to change in the next century, although x^2 was still written xx even by Gauss!

But the clumsy formulas for the cubic and quartic equations were never of much practical use. When a working physicist like Isaac Newton needed to "know" the roots of a polynomial, he developed "Newton's Method" for approximating solutions. The main significance of Ferrari's work was the food for thought (and false hope) it gave to later mathematicians, notably Lagrange.

Exercises on Ferrari's Method

5.23. Explain why it is easy to solve a quartic equation of the form $x^4 + px^2 + r = 0$.

Now let's discuss the special case of a quartic of the form:

$$x^4 + qx + r = 0$$

or

$$(x^2)^2 = -qx - r.$$

Ferrari's trick is to replace x^2 by $x^2 + y$, where y remains to be determined. We get the new equation:

$$(x^2 + y)^2 = 2yx^2 - qx + (y^2 - r).$$

Now if the right-hand side is a perfect square $(Ax + B)^2$, then we can take square roots on both sides and end up with a quadratic equation to solve for x.

5.24. What equation does y have to satisfy in order for the right-hand side to be a perfect square?

5.25. Using this idea for the equation
$$x^4 + 4x - 1 = 0,$$
find a "good" value for y and use it to solve this quartic equation.

5.26. Using calculus, prove that $x^4 - 4qx + r = 0$ has at most two real roots. Find a relationship between q and r that determines whether the number of distinct real roots is 0, 1, or 2. ∎

CHAPTER 6

COMPLEX NUMBERS

And as imagination bodies forth
The forms of things unknown, the poet's pen
Turns them to shapes, and gives to airy nothing
A local habitation and a name

—Wm. Shakespeare

A Midsummer Night's Dream

In the wake of the "casus irreducibilis" connundrum, the mathematical community gradually felt compelled to come to terms with the complex numbers.

The problem, of course, is to give a meaning to the square root of a negative number $-r$. Clearly, we would want:

$$\sqrt{-r} = \sqrt{-1} \cdot \sqrt{r}$$

and so it suffices to give a meaning to $\sqrt{-1}$. Since we are feeling uncomfortable with the whole business we set

$$i = \sqrt{-1}$$

and call i an "imaginary number". Now we want to build the most economical number system (field of numbers) containing all the real numbers plus the new imaginary number i. Since this system should contain "negatives" (additive inverses) for each "number" and since we want to preserve all of the usual rules of arithmetic, we find that we must include a "number" $-i$, and that

$$(-i)^2 = (-1)^2 i^2 = i^2 = -1.$$

Hence, rather than saying that i is "the" square root of -1, it will be more accurate to say that i and $-i$ are the two (imaginary) square roots of -1, i.e. the two (imaginary) solutions of the equation

$$x^2 + 1 = 0.$$

Definition. A complex number is an expression of the form $a+bi$ where a and b are real numbers. Addition is defined by

$$(a+bi) + (c+di) = (a+c) + (b+d)i.$$

Multiplication is defined by

$$(a+bi)(c+di) = (ac-bd) + (ad+bc)i.$$

Addition and multiplication satisfy the extensions of all of the usual associative, commutative, distributive) properties, and additive inverses negatives) exist. So subtractions can be performed.

It is not quite so clear that division by non-zero complex numbers can be performed, but this will follow from Exercise 6.1 below.

Every complex number $\alpha = a+bi$ has a complex conjugate $\bar{\alpha} = a-bi$; a trace:

$$\mathrm{Tr}(\alpha) = \alpha + \bar{\alpha} = 2a,$$

and a norm:

$$N(\alpha) = \alpha \cdot \bar{\alpha} = a^2 + b^2.$$

Exercises

6.1. Prove: If α is a non-zero complex number, then $\frac{1}{\alpha}$ is a complex number; and hence $\frac{\beta}{\alpha}$ is a complex number for every complex number β.

6.2. Prove: Multiplication of complex numbers is commutative.

6.3. Prove: $\overline{\alpha\beta} = \bar{\alpha}\bar{\beta}$.

6.4. Prove: $N(\alpha\beta) = N(\alpha)N(\beta)$. ∎

Notice that the trace and norm of a complex number are both real numbers. In this way division of complex numbers can be reduced to multiplication of complex numbers and division of real numbers by

$$\frac{\alpha}{\gamma} = \frac{\alpha\bar{\gamma}}{N(\gamma)}.$$

With some effort all of the defining properties of a field (of numbers) can be verified for **C**, the complex number field. The comfort level of mathematicians in dealing with complex numbers was increased by the realization around 1800 by several mathematicians (Wessel, Argand, Gauss) that the complex numbers could be identified with the points of the real plane $\mathbf{R}^2$ via the map:

$$a + bi \mapsto (a, b).$$

Then the real numbers live on the x-axis, the "pure" imaginaries live on the y-axis, 0 corresponds to $(0, 0)$, and $N(a + bi)$ is the square of the distance of $a + bi$ from 0, using the usual Euclidean distance on $\mathbf{R}^2$.

Now, as Shakespeare said, we have endowed "airy nothing" with a "local habitation and a name."

Addition of complex numbers corresponds to the additions of vectors in $\mathbf{R}^2$ by the so-called parallelogram rule. But what does multiplication of complex numbers look like geometrically?

This question has a lovely answer, which we shall approach now starting from De Moivre's Formula. Viete's work on the cubic equation may be thought of as establishing a relationship between the problem of solving a cubic equation and the problem of trisecting an angle. Going in the other direction, Leibniz (1675) and De Moivre (1707) found Cardano-type formulas for dividing an angle into n equal parts. This leads to **De Moivre's Formula**:

$$(\cos\theta + i\sin\theta)^n = \cos(n\theta) + i\sin(n\theta).$$

Exercises

6.5. Use the addition formulas for sin and cos to obtain a formula for the product:

$$(\cos\alpha + i\sin\alpha)(\cos\beta + i\sin\beta).$$

6.6. Apply mathematical induction to prove De Moivre's Formula. ∎

Notice that every complex number $z = a + bi$ can be written in "polar" form as

$$z = r(\cos\theta + i\sin\theta)$$

with $r = \sqrt{N(z)}$ and $\tan\theta = b/a$.

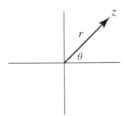

Thus De Moivre's Formula extends to a general formula for complex multiplication:

$$(r(\cos\theta + i\sin\theta))(r'(\cos\theta' + i\sin\theta')) = rr'(\cos(\theta + \theta') + i\sin(\theta + \theta')).$$

In particular notice that complex numbers of norm 1 are precisely the complex numbers on the unit circle S^1 and multiplication by these has a particularly nice interpretation.

Exercises

6.7. Let $\zeta = \cos\theta + i\sin\theta$. Let $\rho_\zeta : \mathbf{C} \to \mathbf{C}$ be the function defined by

$$\rho_\zeta(z) = \zeta \cdot z.$$

Prove: Under the identification of $\mathbf{C}$ with $\mathbf{R}^2$, the map ρ_ζ is identified with counterclockwise rotation about $(0, 0)$ through the angle θ.

6.8a. (Euler) Verify by formal manipulation of power series that

$$\cos z + i \sin z = e^{iz}.$$

6.8b. Define $e^{i\theta} = \cos\theta + i\sin\theta$. Verify that the usual rules for multiplication of exponentials are satisfied.

6.9. Let S^1 denote the set of all complex numbers of norm 1. Prove: S^1 is a group under the operation of complex multiplication. S^1 is called the **circle group**. Can you guess why?

6.10. Prove: The function $\Phi : SO(2) \to S^1$ by

$$\Phi(\rho_\theta) = \cos(\theta) + i\sin(\theta)$$

is an isomorphism of groups.

Definition. Let $(G, \circ)$ and $(H, \cdot)$ be two groups. A function $\Phi : G \to H$ is called a **homomorphism of groups** if

$$\Phi(g \circ g') = \Phi(g) \cdot \Phi(g') \text{ for all } g, g' \in G.$$

(Thus an isomorphism of groups is a special case of a homomorphism of groups. A homomorphism $\Phi : G \to H$ is an isomorphism if and only if Φ is a one-to-one and onto map.)

6.11. (The most famous homomorphism of groups) Prove: The exponential function $E : \mathbf{R} \to S^1$ via

$$E(x) = e^{2\pi i x}$$

is a homomorphism of the group $(\mathbf{R}, +)$ onto the circle group S^1. (It is an example of a covering homomorphism.) Verify that E is not an isomorphism. For which real numbers α is $E(\alpha) = 1$? Prove that this set is a subgroup of $\mathbf{R}$.

Definition. Let $\Phi : G \to H$ be a homomorphism of groups. The kernel of Φ, $Ker(\Phi)$, is the subset of G defined by

$$Ker(\Phi) = \{g \in G : \Phi(g) = e_H\},$$

where e_H is the identity element of the group H.

6.12. Prove: If $\Phi : G \to H$ is a homomorphism of groups, then $Ker(\Phi)$ is a subgroup of G. ∎

De Moivre's Formula can be reversed to give a formula for nth roots of complex numbers. In fact this is the version that De Moivre wrote down:

Let $\alpha = a + bi$ be any complex number. Write

$$\alpha = r(\cos\theta + i\sin\theta)$$

with r a positive real number. Let $r^{1/n}$ be the positive real nth root of r. Then using the fact that cos and sin are periodic of period 2π, we see that the following complex numbers are all nth roots of α:

$$r^{1/n}\left(\cos\frac{\theta}{n} + i\sin\frac{\theta}{n}\right)$$

$$r^{1/n}\left(\cos\frac{\theta + 2\pi}{n} + i\sin\frac{\theta + 2\pi}{n}\right)$$

$$r^{1/n}\left(\cos\frac{\theta + 4\pi}{n} + i\sin\frac{\theta + 4\pi}{n}\right)$$

$$\ldots$$

$$r^{1/n}\left(\cos\frac{\theta + 2(n-1)\pi}{n} + i\sin\frac{\theta + 2(n-1)\pi}{n}\right)$$

More Exercises

6.13. Why do we stop at $(\theta + 2(n-1)\pi)/n$?

6.14. Prove: If α and β are positive real numbers with $\alpha \leq 2\pi$ and $\beta \leq 2\pi$, and if $\cos\alpha = \cos\beta$ and $\sin\alpha = \sin\beta$, then $\alpha = \beta$.

6.15. Draw a picture illustrating the complex number 1 and its five fifth roots.

6.16. Draw a picture illustrating the complex number -8 and its three cube roots.

6.17. Draw a picture illustrating the complex number i and its four fourth roots. ∎

De Moivre's Formula immediately gives us the following amazing fact.

Theorem. *Let α be any nonzero complex number. Then the equation*

$$x^n - \alpha = 0$$

has exactly n distinct roots and all of these roots are complex numbers.

Proof. De Moivre's Formula provides n complex numbers $r_1, \ldots, r_n$, which are roots of $x^n - \alpha$, and these are all distinct by 6.14. By the Factor Theorem

$$x^n - \alpha = (x - r_1)q(x),$$

where $q(x)$ is a polynomial with complex coefficients of degree $n - 1$ and with leading coefficient 1. Then

$$0 = r_i - \alpha = (r_i - r_1)q(r_i)$$

for all $i \geq 2$. Since $r_i - r_1 \neq 0$, we have that $q(r_i) = 0$ for all $i \geq 2$. Proceeding in this manner we obtain

$$x^n - \alpha = (x - r_1)(x - r_2)\cdots(x - r_n).$$

As in the proof of the Corollary to the Factor Theorem, we conclude that $\{r_1, r_2, \ldots, r_n\}$ is the set of all roots of $x^n - \alpha$. ∎

The amazing fact is that we introduced one "imaginary" symbol i in order to solve one quadratic equation:
$$x^2 + 1 = 0.$$
Because of completing the square trick, it is not so surprising that *every* quadratic equation
$$x^2 + bx + c = 0$$
can be solved using only numbers of the form $a + di$ with a and d real numbers. But it is truly amazing that we do not need to introduce *new* imaginary symbols to solve equations like
$$x^4 + 1 = 0$$
or
$$x^6 + 3 = 0.$$
This raises the question: Do we need to introduce new imaginary symbols in order to solve *any* polynomial equations? Or do the complex numbers suffice? This was a hot topic of debate in the early 18th century. Leibniz thought that the complex numbers did not suffice. Others disagreed. Of course if formulas like Cardano's Formulas could be found to solve polynomial equations of all possible degrees, then the solution of polynomial equations would be reduced to the operations of addition, subtraction, multiplication, division, and the extraction of roots. In this case De Moivre's Formula would indeed guarantee that the complex numbers would suffice to solve every polynomial equation. But no Cardano-type formulas were known for equations of degree greater than 4. In the next chapter, we shall see how this debate evolved. But first some more exercises.

More Exercises

6.18. What is the complex conjugate of the complex number $re^{i\theta}$? Illustrate with a picture.

6.19a. Prove that if α is an nth root of a real number a, then $\bar{\alpha}$ is also an nth root of a.

6.19b. More generally prove that if α is a root of a polynomial $p(x)$ which has real coefficients, then $\bar{\alpha}$ is also a root of $p(x)$.

6.20. Prove that if a is a real number, then the polynomial $x^n - a$ can be factored as a product of real polynomials of degree 1 and 2.

6.21. Factor the polynomials $x^2 - 1$, $x^3 - 1$, $x^4 - 1$, $x^6 - 1$, $x^8 - 1$, and $x^{12} - 1$ as a product of "irreducible" real polynomials of degree 1 and 2.

*6.22. Challenge problem: Factor $x^5 - 1$ as a product of real polynomials of degree 1 and 2. (*Hint*: Show that if α is a root, then $Tr(\alpha) = \alpha + (1/\alpha)$. Manipulate the equation $(x^5 - 1)/(x - 1)$ to find a quadratic equation satisfied by $t = Tr(\alpha)$ and solve it to find t. Now write down a real quadratic polynomial having α as a root.) ∎

Notice that the n complex roots of $x^n - 1$ lie at the vertices of a regular n-gon (polygon with n sides) for $n \geq 3$. Draw some illustrative pictures for small values of n.

Armed with De Moivre's Formula, we can now reexamine the Casus Irreducibilis by Cardano's Method.

Exercises (The Cubic and Quadratic Equations Revisited)

6.23a. Find all solutions of $x^3 - 6x + 5 = 0$ again, this time using Cardano's Method. Draw a picture illustrating u^3 and its three cube roots. Compare with the solutions you obtained before.

6.23b. Do the same for $x^3 - 6x + 4 = 0$.

6.23c. Do the same for $x^3 - 3x - 1 = 0$. Compare with your solution via Viete's Method.

6.23d. Do the same for $x^3 - 12x - 8 = 0$.

6.24a. Derive the quadratic formula for the solutions of $x^2 - bx + c = 0$, where b and c are arbitrary complex numbers. Verify that it looks exactly the same as the usual quadratic formula and that the solutions are complex numbers.

6.24b. Solve the equation $x^2 - ix + 1 = 0$.

6.24c. Solve the equation $x^2 - 2x + i = 0$. Draw an illustrative picture showing the roots. ∎

It will be of considerable interest to us later to study a subset of the complex numbers that plays a role analogous to that of the ordinary integers in the real numbers. Gauss first distinguished this set and so we call it the **Gaussian integers**:

$$\mathbf{Z}[i] = \{a + bi : a, b \in \mathbf{Z}\}.$$

Exercises on the Gaussian Integers

6.25. Prove that the Gaussian integers form a ring of numbers, i.e., that $\mathbf{Z}[i]$ is closed under addition, subtraction, and multiplication.

6.26. Prove that the norm of a Gaussian integer is a nonnegative (ordinary) integer and is 0 if and only if the Gaussian integer is 0.

6.27. List all Gaussian integers of norm less than 14. Draw a picture illustrating their location in the Argand plane representation of the complex numbers. Does every nonnegative integer occur as the norm of a Gaussian integer? ∎

Exercises on the complex numbers as matrices. *Matrix algebra was a later historical development of the 19th century. Peeking ahead we see one more way of thinking about the complex numbers.*

6.28. Identify the complex numbers $\alpha = a + bi$ and $\gamma = c + di$ with the 2×2 matrices of real numbers

$$f(\alpha) = \begin{bmatrix} a & b \\ -b & a \end{bmatrix}$$

and

$$f(\gamma) = \begin{bmatrix} c & d \\ -d & c \end{bmatrix}.$$

Prove:

$$f(\alpha + \gamma) = f(\alpha) + f(\gamma)$$

and
$$f(\alpha \cdot \gamma) = f(\alpha)f(\gamma),$$
where the operations on the right-hand side are the usual matrix addition and multiplication. ∎

Remark. *In fact we may extend the concept of a ring of numbers to a more abstract concept of a* **ring** *R as a set of objects closed under two binary operations* $+$ *and* $\cdot$ *such that* $(R, +)$ *is an abelian group,* $\cdot$ *is an associative (but not necessarily commutative) operation, and the following distributive laws hold:*
$$a \cdot (b + c) = a \cdot b + a \cdot c$$
and
$$(a + b) \cdot c = a \cdot c + b \cdot c$$
for all $a, b, c \in R$. *Note that both distributive laws must be specified because the multiplication operation* $\cdot$ *is not assumed to be commutative.*

Given two rings $(R, +, \cdot)$ and $(S, +, \circ)$, we say that a function $f : R \to S$ is a **homomorphism of rings** if
$$f(r + r') = f(r) + f(r')$$
and
$$f(r \cdot r') = f(r) \circ f(r')$$
for all $r, r' \in R$.

In this language the set $M_2(\mathbf{R})$ of all 2×2 matrices of real numbers is a ring and the function $f : \mathbf{C} \to M_2(\mathbf{R})$ defined in 6.28 is a homomorphism of rings. It is injective but not surjective, hence not an isomorphism of rings. Indeed the commutative law for multiplication holds in $\mathbf{C}$ but not in $M_2(\mathbf{R})$.

6.29a. Prove: The set $2\mathbf{Z}$ of even integers is a ring under ordinary addition and multiplication.

6.29b. Prove: The function $f : \mathbf{Z} \to 2\mathbf{Z}$ by $f(n) = 2n$ is an isomorphism of groups but is not a homomorphism of rings.

6.30. Prove: $N(\alpha) = det(f(\alpha))$ and $Tr(\alpha) = Tr(f(\alpha))$, where the trace $Tr(A)$ of a matrix A is defined to be the sum of the entries on the diagonal of A.

6.31. Prove: If α is a complex number of norm 1, then $f(\alpha)$ is a rotation matrix. In fact f defines an isomorphism between the groups S^1 and the group $SO(2)$. (This is basically the same isomorphism as the one given in 6.10.)

6.32. Let $R(x, y) = (x, -y)$ be the reflection across the x-axis. Write its matrix with respect to the "standard" basis $\{(1, 0), (0, 1)\}$ for $\mathbf{R}^2$. If you think of R as a function on the complex numbers, what function is it? ∎

Remark. *The polar decomposition* $\alpha = re^{i\theta}$ *of complex numbers, when considered in the context of real* 2×2 *matrices, has a natural extension to a general* *polar decomposition of real $n \times n$ matrices as the product $A = RU$ where U is an $n \times n$ orthogonal matrix (i.e., $UU^t = I$) and R is a symmetric positive semidefinite matrix (i.e., $R = R^t$ and the eigenvalues of R are nonnegative real numbers). Note that all eigenvalues of U are of the form $e^{i\theta}$. This polar decomposition follows from the Spectral Theorem applied to the symmetric matrix AA^t. It generalizes to complex $n \times n$ matrices A with U then a unitary matrix.*

CHAPTER 7

SYMMETRIC POLYNOMIALS AND THE FUNDAMENTAL THEOREM OF ALGEBRA

Symmetry seems to be absolutely fascinating to the human mind.

–Richard Feynman

The existence of exactly n complex nth roots for any number lent support to the idea that every polynomial equation of degree n should have *exactly n* complex solutions (counting multiplicity). A (somewhat cryptic) version of this statement was first made (without proof) by Girard as early as 1629 (*before* Descartes published the Factor Theorem!), along with formulas expressing the coefficients as symmetric functions of the roots. Nevertheless as late as the early 1700s, Leibniz thought he had a counterexample. D'Alembert published a (somewhat) incomplete proof of this " Fundamental Theorem of Algebra" using the methods of calculus in 1746. Euler attempted a more algebraic proof in 1749, which was improved by de Foncenex and then Lagrange in 1772.

There was one big problem with the Euler–Foncenex–Lagrange (EFL) proof, and this was pointed out by Gauss, who proposed his own calculus-based proof in 1799 and a second more algebraic proof in 1816. In spite of the "fatal flaw" in the EFL proof (which we shall explain later in this section), we shall present it in this section. We will in fact fix the fatal flaw in Chapter 19 using an idea of Galois (1830).

First we need to discuss symmetric polynomials.

Definition. A polynomial of n variables $f(r_1, r_2, \ldots, r_n)$ is called a **symmetric polynomial** if it remains unchanged after any permutation of the variables.

Thus
$$f(r_1, r_2, \ldots, r_n) = r_1 + r_2 + \cdots + r_n$$
and
$$g(r_1, r_2, \ldots, r_n) = r_1 r_2 \ldots r_n$$
are symmetric polynomials, but the polynomial
$$h(r_1, r_2, \ldots, r_n) = r_1 - r_2$$
is not symmetric since the permutation σ given by
$$\sigma = \begin{bmatrix} r_1 & r_2 & \cdots & r_n \\ r_2 & r_1 & \cdots & r_n \end{bmatrix}$$
satisfies $\sigma(h) = k$, where k is the polynomial
$$k(r_1, r_2, \ldots, r_n) = r_2 - r_1.$$

Symmetric polynomials in n variables enter the theory of polynomial equations in one variable in the following way, as observed by Girard. If
$$p(x) = x^n + a_1 x^{n-1} + \cdots + a_{n-1} x + a_n$$
is a polynomial whose roots are $r_1, r_2, \ldots, r_n$, then by Descartes' Factor Theorem we have
$$x^n + a_1 x^{n-1} + \cdots + a_{n-1} x + a_n = (x - r_1)(x - r_2) \cdots (x - r_n).$$

Equating coefficients we get each of the a_i's expressed as a symmetric function of the r_i's:
$$-a_1 = r_1 + r_2 + \cdots + r_n,$$
$$a_2 = r_1 r_2 + r_1 r_3 + \cdots + r_1 r_n + r_2 r_3 + \cdots + r_{n-1} r_n,$$
and so on to
$$(-1)^n a_n = r_1 r_2 \ldots r_n.$$

The n expressions on the right are called the **elementary symmetric polynomials** of the r_i's: $s_1, s_2, \ldots, s_n$. Obviously not every symmetric polynomial is an elementary symmetric polynomial. However the following theorem is fundamental and was probably known to Newton, though first explicitly proved somewhat later by Waring.

Theorem. *Every symmetric polynomial $f(r_1, r_2, \ldots, r_n)$ is expressible as a polynomial in the elementary symmetric polynomials; i.e., there exists a polynomial $F(x_1, x_2, \ldots, x_n)$ depending on f such that*
$$f(r_1, r_2, \ldots, r_n) = F(s_1, s_2, \ldots, s_n).$$

The proof amounts to an algorithm for constructing F from f, and rather than describe it "abstractly," let's look at an illustrative example in three variables. To save ink, let's call the variables r, s, and t instead of r_1, r_2, and r_3.

If we want to cook up an example of a symmetric polynomial in, say, three variables, we can "symmetrize" any monomial by adding up all of its possible permutations. For example starting with the monomial:

$$m(r, s, t) = r^2 s,$$

we get the symmetrized polynomial

$$p(r, s, t) = r^2 s + r^2 t + s^2 r + s^2 t + t^2 r + t^2 s.$$

An important feature of any algorithm is to have an order of operation so that you know you are making steady progress in the right direction, rather than going around in circles. In order to do this we need to have a way to say that the polynomial $p(r, s, t)$ is "bigger" than some other polynomial $q(r, s, t)$. Once we have that, we will look for an algorithm to make $p(r, s, t)$ systematically "smaller and smaller" until we finally reach 0.

For a monomial $f(r, s, t) = a r^i s^j t^k$, we call its degree vector (i, j, k). Thus our $m(r, s, t)$ has degree vector $(2, 1, 0)$. We order the degree vectors lexicographically reading from left to right. Thus $(3, 0, 0) > (2, 1, 1) > (2, 1, 0) > (2, 0, 7) > (0, 3, 5)$, for example. Thus the highest term of our polynomial $p(r, s, t)$ is $r^2 s$, which has degree vector $(2, 1, 0)$. We order our polynomials according to the size of their highest terms.

Here is a crucial point: If $f(r, s, t)$ is a symmetric polynomial containing a monomial with degree vector (i, j, k), then by symmetry, f must also contain monomials with degree vector every possible permutation of (i, j, k). In one of these, $i \geq j \geq k$. In particular, the highest term of f has degree vector (a, b, c) with $a \geq b \geq c$, by the nature of our ordering.

Here is another crucial point: The highest terms of the elementary symmetric polynomials in r, s, t have degree vectors $(1, 0, 0)$, $(1, 1, 0)$ and $(1, 1, 1)$. When we multiply monomials, we add their degree vectors, so we can build a "monomial" in s_1, s_2, s_3 having any highest term $m(r, s, t)$, so long as the degree vector (i, j, k) for $m(r, s, t)$ is nondecreasing, i.e. $i \geq j \geq k$. For example if we want the degree vector to be $(2, 1, 0)$, we note that

$$(2, 1, 0) = (1, 0, 0) + (1, 1, 0)$$

and so $s_1 s_2$ should do the job. Let's check.

$$s_1 s_2 = (r + s + t)(rs + rt + st) = r^2 s + r^2 t + rt^2 + s^2 t + rs^2 + st^2 + 3rst.$$

The point is that $p(r, s, t) - s_1 s_2$ has a smaller highest degree vector than $p(r, s, t)$. So we are making progress! In fact this example was too easy:

$$p(r, s, t) - s_1 s_2 = -3rst = -3s_3$$

and so we are done:

$$p(r, s, t) = s_1 s_2 - 3s_3.$$

Of course in general the procedure takes much longer but clearly we keep simplifying the problem, so like Gaussian elimination, we must eventually succeed. You can try a few harder examples in the exercises.

Exercises

7.1. Express $r^2 + s^2 + t^2$ as a polynomial in the elementary symmetric polynomials $s_1 = r + s + t$, $s_2 = rs + rt + st$, and $s_3 = rst$.

7.2. Do the same for $r^2 + s^2 + t^2 + u^2$ in terms of the four elementary symmetric polynomials in the variables $r, s, t,$ and u.

7.3. Do the same for $r^3 + s^3 + t^3$.

7.4. (*Bonus*) Let $p_k = r_1^k + r_2^k + \cdots + r_n^k$, $k = 1, 2, \ldots, n$. Prove Newton's Identities:

$$p_k - s_1 p_{k-1} + s_2 p_{k-2} - \cdots + (-1)^{k-1} s_{k-1} p_1 + (-1)^k k s_k = 0. \blacksquare$$

THE EFL PROOF OF THE FUNDAMENTAL THEOREM OF ALGEBRA

We now present the Euler–de Foncenex–Lagrange proof of the Fundamental Theorem of Algebra. It rests on the assumption (challenged by Gauss, but justified later by Galois and Kronecker) that any polynomial with real coefficients has n roots in *some* field of numbers possibly *larger* than the complex number field. In other words Euler took for granted that you could always add more imaginary symbols to get a field of numbers big enough to contain all of the roots of a given polynomial $p(x)$. Specifically he took for granted that this *set* of imaginary symbols could be enclosed in a *field* that really satisfied all the usual associative, commutative, and distributive laws. His task, as he saw it, was then to prove that this *big field* was really just the complex numbers after all. Gauss questioned whether this *big field* could really be constructed at all. Galois and Kronecker were the first to prove that it could.

Assuming this for now, we sketch the EFL proof.

Theorem. *The following two statements are equivalent. (The first is usually called the Fundamental Theorem of Algebra.)*

1. *For all real polynomials $p(x)$ of degree $n \geq 1$, $p(x) = 0$ has n solutions in* **C**.
2. *For all real polynomials $p(x)$ of degree $n \geq 1$, $p(x) = 0$ has at least one solution in* **C**.

Proof. Clearly (1) implies (2). We prove that (2) implies (1) by induction on the degree n. If $n = 1$, then (2) obviously implies (1). Suppose then that (2) holds and (1) holds for all polynomials of degree less than n. Let $p(x)$ be a polynomial of degree n with real coefficients. By (2) $p(\alpha) = 0$ for some complex number α. Then also $p(\overline{\alpha}) = 0$ by 6.19b. And so by the Factor Theorem

$$p(x) = (x - \alpha)q(x) \text{ if } \alpha \in \mathbf{R}$$

and if $\alpha \neq \overline{\alpha}$, then

$$p(x) = (x^2 - Tr(\alpha)x + N(\alpha))q(x)$$

with $q(x)$ a real polynomial in either case. By induction the equation $q(x) = 0$ has $deg(q)$ solutions in **C** and so $f(x) = 0$ has n solutions in **C**, as desired. $\blacksquare$

No proof of the Fundamental Theorem of Algebra is purely algebraic. The point in the EFL proof that is not algebraic is the following appeal to calculus.

Theorem. *Let $p(x)$ be a polynomial with real coefficients of odd degree. Then $p(x)$ has at least one real root.*

Sketch of a Proof. We may assume that the leading coefficient of $p(x)$ is positive. Then for "sufficiently large" positive values of x, $p(x) > 0$ and for "sufficiently negative" values of x, $p(x) < 0$. Therefore by the Intermediate Value Theorem, there is some real number r for which $p(r) = 0$; i.e., r is a real root of $p(x)$. ∎

We also need the following more elementary fact, which you proved in 6.24a.

Theorem. *Let $q(x) = x^2 - \gamma x + \delta$ be a quadratic polynomial with complex coefficients. Then $q(x)$ has two complex roots.*

Now we can state and prove the Fundamental Theorem of Algebra.

Fundamental Theorem of Algebra. *Let $p(x)$ be any polynomial of degree $n \geq 1$ with real coefficients. Then $p(x)$ has n roots in $\mathbf{C}$.*

As we saw earlier, it will suffice to prove the following theorem.

Theorem. *Let $p(x)$ be a polynomial of degree $n \geq 1$ with real coefficients. Then $p(x)$ has at least one root in $\mathbf{C}$.*

Proof. For any positive integer n, we define $e_2(n) = e$ to be the nonnegative integer e such that $n = 2^e n_0$ for some odd number n_0. If $p(x)$ is a polynomial of degree n, we define $e_2(p) = e_2(n)$. We shall prove the theorem by induction on $e_2(p)$ not on the degree of p. In fact we shall "reduce" to a family of polynomials g_c of much larger degree than $p(x)$ but with $e_2(g_c) = e_2(p) - 1$ for all c.

To begin the induction we treat the case $e_2(p) = 0$. But this is the case where p has odd degree and then by calculus we know that p has a root in $\mathbf{R}$ (as discussed earlier).

Now we do the inductive step. Thus we assume that the theorem is true whenever $e_2(p) = m - 1$ and treat the case when $e_2(p) = m \geq 1$. We let $r_1, r_2, \ldots, r_n$ be the roots of $p(x)$ and for *each* positive integer c we form the set of "numbers"

$$y_{ij}(c) = r_i + r_j + c r_i r_j,$$

where $1 \leq i < j \leq n$. Thus for each c we have $n(n-1)/2$ numbers $y_{ij}(c)$ and a key point of the argument is that since n is even, $n(n-1)/2$ is divisible by 2^{m-1} but *not* by 2^m.

For each c, construct the monic polynomial g_c whose roots are precisely the $n(n-1)/2$ numbers $y_{ij}(c)$. The coefficients of g_c are (up to sign) the elementary symmetric polynomials in the $y_{ij}(c)$'s and so they are symmetric polynomials in the r_i's.

(Though intuitively clear, this is a somewhat delicate point. A rigorous proof would go as follows: Let σ be any permutation of $\{r_1, r_2, \ldots, r_n\}$ (and of $\{1, 2, \ldots, n\}$). Then

$$\sigma(y_{ij}(c)) = y_{\sigma(i)\sigma(j)}(c)$$

and so σ permutes the roots of g_c. Then in an obvious sense, $\sigma(g_c) = g_c$ and so σ maps each coefficient $a_k(r_1, \ldots, r_n)$ of g_c to itself. Hence each $a_k(r_1, \ldots, r_n)$ is a symmetric polynomial in the r_i's, as claimed.)

Since the $y_{ij}(c)$'s are symmetric polynomials in the r_i's, they are polynomials in the elementary symmetric polynomials in the r_i's by the Fundamental Theorem on Symmetric Polynomials. But the elementary symmetric polynomials in the r_i's are (up to sign) the *real* coefficients of $p(x)$. So the coefficients of each of the polynomials $g_c(x)$ are *real numbers*. Also

$$e_2(g_c) = e_2\left(\frac{n(n-1)}{2}\right) = m - 1.$$

Thus our inductive hypothesis applies and tells us this:

For each c, g_c has a root in **C**.

Thus for each positive integer c, there exists a pair (i, j) such that

$$y_{ij}(c) = r_i + r_j + cr_ir_j \text{ is a complex number.}$$

Since there are infinitely many natural numbers, there must exist two *different* natural numbers c and d for which the *same pair* (i, j) "works"; that is,

$$r_i + r_j + cr_ir_j = \alpha \in \mathbf{C}$$

and

$$r_i + r_j + dr_ir_j = \beta \in \mathbf{C}.$$

Subtracting and dividing by $c - d$ (which is not 0), we get that $r_ir_j = \delta \in \mathbf{C}$. Then also

$$r_i + r_j = \alpha - c\delta = \gamma \in \mathbf{C}.$$

Now r_i and r_j are the roots of the quadratic equation

$$x^2 - \gamma x + \delta = 0$$

whose coefficients are in **C**. We conclude that both r_i and r_j are complex numbers, and we are done! ∎

Definition. A polynomial $p(x)$ with coefficients in a field of numbers F (such as **Q** or **R** or **C**) is said to be reducible in $F[x]$ if there exist **nonconstant** polynomials $g(x)$ and $h(x)$ with coefficients in F such that

$$f(x) = g(x)h(x).$$

Otherwise $p(x)$ is said to be irreducible in $F[x]$.

Notice that the concept of an irreducible polynomial depends *fundamentally* on the field of coefficients that are allowed. By the Fundamental Theorem of Algebra the only nonconstant irreducible polynomials in **C**[x] are the polynomials of degree 1: $p(x) = ax + b$. On the other hand there are *lots* of irreducible polynomials in **Q**[x].

Exercises

7.5a. Using the Fundamental Theorem of Algebra, prove that every polynomial with real coefficients can be factored into a product of polynomials of degree 1 and 2, each of which has real coefficients. (*Note*: This fact is crucial to the partial fractions method for computing the integral of an arbitrary rational function $p(x)/q(x)$.

In fact this was one of the main motivations in the 18th century for proving the Fundamental Theorem.)

7.5b. Describe precisely which are the irreducible polynomials in $\mathbf{R}[x]$.

7.5c. Give an example of a quadratic polynomial that is irreducible in $\mathbf{Q}[x]$ but not in $\mathbf{R}[x]$. Do the same for a cubic polynomial.

7.6a. Prove: Counting multiplicity, a polynomial of even degree with real coefficients always has an even number (possibly 0) of real roots.

7.6b. Give an example of a quartic polynomial with real coefficients and no real roots. Factor it into two irreducible quadratic polynomials with real coefficients. ∎

LAGRANGE'S 1770–1771 PAPER

Now let's rethink the solution of the quadratic and cubic equations as Lagrange did. First consider the quadratic equation

$$x^2 + bx + c = 0.$$

Let's call the roots (which we want to find) r and s. Then by the Factor Theorem

$$x^2 + bx + c = (x - r)(x - s) = x^2 - (r + s)x + rs.$$

So we *know* the symmetric functions $r + s$ and rs. We want to find the asymmetric functions r and s. This is of course the original Babylonian version of the quadratic equation problem, "jazzed up" in the language of symmetric functions.

Lagrange talks about how many "values" (valeurs in French) a function takes when you permute the variables. Of course there is only one nontrivial permutation of the set $\{r, s\}$:

$$\sigma = \begin{bmatrix} r & s \\ s & r \end{bmatrix},$$

Now we may think of σ as a permutation of the polynomials in r and s by the obvious extension

$$\sigma(cr^i s^j) = cs^i r^j \text{ and } \sigma(f + g) = \sigma(f) + \sigma(g),$$

where f and g are any two polynomials in the variables r and s, and c is any constant.

Thus symmetric polynomials like $r + s$ and rs take only one value; i.e.,

$$\sigma(r + s) = r + s$$

and

$$\sigma(rs) = rs,$$

but the polynomial r takes *two* values: r and s. So does the polynomial s. Also the polynomial δ defined by

$$\delta(r, s) = r - s$$

takes two values, but δ is "better" than r and s because

$$\sigma(\delta) = \sigma(r - s) = s - r = -\delta,$$

and so the polynomial δ^2 takes only *one* value; i.e., it is symmetric:

$$\sigma(\delta^2) = (\sigma(\delta))^2 = (-\delta)^2 = \delta^2.$$

But then δ^2 can be expressed as a polynomial in the elementary symmetric polynomials. Indeed:

$$\delta^2 = (r-s)^2 = (r+s)^2 - 4rs = b^2 - 4c.$$

And now we can find δ by taking square roots and from δ and the symmetric polynomial

$$s_1(r, s) = r + s = -b,$$

we easily find r and s. Of course this was much ado about very little. We just rediscovered the quadratic formula the hard way.

Now let's try to apply the same reasoning to the cubic equation

$$x^3 - px + q = 0.$$

Again set

$$x^3 - px + q = (x-r)(x-s)(x-t).$$

Here the process is more complicated. Lagrange's new idea is to consider the "Lagrange resolvent"

$$\lambda(r, s, t) = r + \omega s + \omega^2 t,$$

where $1, \omega,$ and ω^2 are the complex cube roots of 1. Now λ takes *six* values under permutation of $r, s,$ and t. (Ah, so that's why Cardano ended up with an equation of degree 6. Remember?) But it is easy to see that if we set

$$\mu(r, s, t) = s + \omega r + \omega^2 t,$$

then the six values of λ are

$$\lambda, \omega\lambda, \omega^2\lambda, \mu, \omega\mu, \omega^2\mu$$

and so, since $\omega^3 = (\omega^2)^3 = 1$, the function λ^3 takes only *two* values under permutation:

$$\lambda^3 \text{ and } \mu^3.$$

Since λ^3 and μ^3 are the two values taken by λ^3 under permutation, it follows that

$$\lambda^3 + \mu^3 \text{ and } \lambda^3\mu^3$$

are symmetric polynomials of r, s and t. But this means that the quadratic equation

$$x^2 - (\lambda^3 + \mu^3)x + (\lambda^3\mu^3)$$

has coefficients expressible in terms of the coefficients p and q of our original cubic and its roots are λ^3 and μ^3. So we can find λ^3 and μ^3, and hence by taking cube roots we can find λ and μ. Then we have a system of linear equations:

$$r + \omega s + \omega^2 t = \lambda$$
$$\omega r + s + \omega^2 t = \mu$$
$$r + s + t = 0$$

to solve in order to find r, s, and t.

Lagrange extends these ideas even further to explain the solution of the quartic equation, but he can't make the method work for equations of degree higher than 4. In this sense his paper was a grand *failure*. But it was a failure only in the sense that Columbus' voyages were a failure. He had touched upon a *new world*: the world of groups.

Exercises

Recall that the symmetric group S_n is the group of all permutations of the set $\{1, 2, \ldots, n\}$. Let $f = f(r_1, r_2, \ldots, r_n)$ be a polynomial. Define

$$\sigma(f) = f(r_{\sigma(1)}, \quad r_{\sigma(2)}, \ldots, r_{\sigma(n)}).$$

For example if

$$f = f(r_1, r_2, r_3) = r_1^3 + r_2 r_3,$$

and if

$$\sigma = \begin{bmatrix} 1 & 2 & 3 \\ 2 & 3 & 1 \end{bmatrix},$$

then

$$\sigma(f) = r_2^3 + r_3 r_1 = r_2^3 + r_1 r_3. \quad \blacksquare$$

Definition. If $f = f(r_1, r_2, \ldots, r_n)$, then

$$(S_n)_f = \{\sigma \in S_n : \sigma(f) = f\};$$

i.e., $(S_n)_f$ is the set of all permutations in S_n that fix f. We may think of $(S_n)_f$ as the symmetry group of the polynomial f.

7.7. Prove. For every polynomial $f = f(r_1, r_2, \ldots, r_n)$, $(S_n)_f$ is a subgroup of S_n. (*Hint*: This is very easy and does not depend on knowing *anything* about f.) $\blacksquare$

Let's call a polynomial an **alternating polynomial** if it takes exactly two values under permutation of the variables. The **discriminant** $\delta(r_1, r_2, \ldots, r_n)$ is defined by

$$\delta(r_1, r_2, \ldots, r_n) = \prod_{i<j}(r_i - r_j).$$

Thus when $n = 3$, we have

$$\delta(r_1, r_2, r_3) = (r_1 - r_2)(r_1 - r_3)(r_2 - r_3).$$

7.8. Prove: For every $\sigma \in S_n$, either $\sigma(\delta) = \delta$ or $\sigma(\delta) = -\delta$. Thus δ is an alternating polynomial and δ^2 is a symmetric polynomial.

SYMMETRIC POLYNOMIALS AND THE FUNDAMENTAL THEOREM OF ALGEBRA   77

Remark. *It's not pretty but with patience using Waring's method you can express δ^2 in terms of the elementary symmetric polynomials.*

7.9. Check: When $n = 3$ and
$$(x - r_1)(x - r_2)(x - r_3) = x^3 - px + q,$$
we have $\delta^2 = 4p^3 - 27q^2$.

Definition. The **alternating group** A_n is the subgroup $(S_n)_\delta$ of all permutations in S_n fixing the discriminant function δ.

It is not obvious, but $A_n = (S_n)_g$ for any alternating function g.

We will now look at the case $n = 3$. The number 3 is not important. It is just small enough so that we can do examples and big enough so that the examples are not *too* easy.

Exercises

7.10a. Write down every permutation in S_3.

7.10b. Now write down *every* subgroup of S_3.

7.10c. For each subgroup H of S_3, write down a polynomial $f = f(r_1, r_2, r_3)$ such that $H = (S_3)_f$.

7.11. Following Lagrange, if $f = f(r_1, r_2, r_3)$ is a polynomial and σ is a permutation in S_3, we call $\sigma(f)$ a **value** of f. (*Note*: A value of f is another polynomial $g = g(r_1, r_2, r_3)$.) What is the maximum possible number of different values of f? If possible give examples of functions $f = f(r, s, t)$ that have exactly 1, 2, 3, 4, ..., (up to the maximum) different values. For each example f, write down the group $(S_3)_f$ of all permutations that fix f. Do you see any connection between the cardinality $|(S_3)_f|$ of the set $(S_3)_f$ and the number of different values of f?

7.12. Let H be a subgroup of S_n. Suppose that
$$H = \{\sigma_1, \sigma_2, \ldots, \sigma_m\}.$$
If $f = f(r_1, r_2, \ldots, r_n)$ is a polynomial, let
$$\bar{f} = \sum_{i=1}^{m} \sigma_i(f).$$
Prove: $H \leq (S_n)_{\bar{f}}$; i.e., $\sigma(\bar{f}) = \bar{f}$ for all $\sigma \in H$.

7.13. (Bonus) Prove the following theorem of Mathieu: If H is any subgroup of S_n, there is a function $f(r_1, r_2, \ldots, r_n)$ such that $(S_n)_f = H$. ∎

CHAPTER 8

PERMUTATIONS AND LAGRANGE'S THEOREM

The bad (or good, depending on your taste) news is that $n!$ gets large very quickly. So it already becomes unpleasant to write down all the permutations in S_4, the group of all permutations of $\{1, 2, 3, 4\}$. It helps to have better notation, which was developed in the years after Lagrange's paper by Ruffini and Abbati and finally perfected by Cauchy around 1815.

As an example, for variety, let's look at a geometrical permutation group, or symmetry group. Specifically, let's consider the symmetries of the regular hexagon illustrated here.

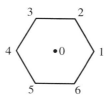

Let ρ denote the 60° counterclockwise rotation of the plane about the point 0. As a permutation of the vertices of our hexagon,

$$\rho = \begin{pmatrix} 1 & 2 & 3 & 4 & 5 & 6 \\ 2 & 3 & 4 & 5 & 6 & 1 \end{pmatrix}.$$

We may think of ρ as inducing a flow on the points of the plane, and specifically on the set $V = \{1, 2, 3, 4, 5, 6\}$, pushing 1 to $\rho(1) = 2$, 2 to $\rho(2) = 3$, etc. Indeed the vertices seem to be orbiting about the central point 0, which remains fixed. In the spirit of this language of dynamical systems we make the following definition:

Definition. Let σ be a permutation of the set $S = \{a, b, c, \dots\}$. The σ-**orbit** on S containing a is the set $a^\sigma = \{a, \sigma(a), \sigma^2(a), \dots\}$.

78

Now the set a^σ looks infinite, but if S is a finite set or σ is a permutation of finite order, then there must be repetitions in the infinite list. Assume henceforth that the set S is finite. Let $\sigma^i(a)$ and $\sigma^{i+r}(a)$ be the closest pair of repeated letters with $r > 0$. Then

$$\sigma^{i+r}(a) = \sigma^i(\sigma^r(a)) = \sigma^i(a).$$

Let $\sigma^r(a) = b$. Then our equation says

$$\sigma^i(b) = \sigma^i(a).$$

But permutations are one-to-one functions. So this means $b = a$; i.e., a gets repeated later in the list. As soon as we have $\sigma^r(a) = a$, we obviously get

$$\sigma^{r+1}(a) = \sigma(a), \sigma^{r+2}(a) = \sigma^2(a), \ldots;$$

i.e., we keep cycling through the same points over and over again. Thus we have

$$a^\sigma = \{a, \sigma(a), \ldots, \sigma^{r-1}(a)\}.$$

Moreover there are no repetitions on this list, since we chose r to be the smallest distance between repetitions.

Going back to our example of the rotation ρ of the vertices of the hexagon, we find that this smallest integer r is 6:

$$\rho(1) = 2, \rho^2(1) = 3, \ldots, \rho^5(1) = 6, \rho^6(1) = 1,$$
$$\rho^7(1) = \rho(1) = 2, \rho^8(1) = \rho^2(1) = 3, \text{ etc.}$$

Now if we take any other point i in the ρ-orbit of 1 and start "pushing" it with ρ, we obviously again run through the same orbit, though in a different order, for example

$$\rho(2) = 3, \rho^2(2) = 4, \ldots, \rho^5(2) = 1, \rho^6(2) = 2.$$

It's like watching the hours on a clock, but starting at a different hour. Thus returning to the set S and the permutation σ we have

$$a^\sigma = b^\sigma \text{ for all } b \in a^\sigma.$$

Thus a σ-orbit is *uniquely determined* by *any* one of its members.

Theorem. *If a and b are in S, then either*

(i) $a^\sigma = b^\sigma$ *or*
(ii) $a^\sigma \cap b^\sigma = \emptyset$.

Proof. Suppose $a^\sigma \cap b^\sigma \neq \emptyset$. Let c be a point in the intersection. Since $c \in a^\sigma$, we have

$$a^\sigma = c^\sigma.$$

On the other hand, since $c \in b^\sigma$, we have

$$b^\sigma = c^\sigma.$$

Thus $a^\sigma = b^\sigma$. ∎

This shows that the σ-orbits on S form a *disjoint partition* of S. Equivalently we may say that the property of being in the same σ-orbit defines an equivalence relation on the set S.

In the case of our example of the rotation ρ, the partition of V into ρ-orbits is rather boring. There is only one ρ-orbit: V. But if we consider powers of ρ, the partitions become more interesting. Thus ρ^2 is the 120° counterclockwise rotation of the plane and this time there are two ρ^2-orbits on V:

$$1^{\rho^2} = \{1, 3, 4\} \text{ and } 2^{\rho^2} = \{2, 4, 6\}.$$

Note that $1^{\rho^2} = 3^{\rho^2} = 5^{\rho^2}$ and $2^{\rho^2} = 4^{\rho^2} = 6^{\rho^2}$.

And $V = \{1, 3, 5\} \cup \{2, 4, 6\}$ as a union of ρ^2-orbits. Likewise ρ^3 is the 180° rotation of the plane about 0 and there are three ρ^3-orbits on V:

$$1^{\rho^3} = \{1, 4\},\ 2^{\rho^3} = \{2, 5\},\ \text{and}\ 3^{\rho^3} = \{3, 6\}.$$

Again $V = \{1, 4\} \cup \{2, 5\} \cup \{3, 6\}$ as a union of ρ^3-orbits.

Correspondingly we can completely describe the permutation σ by giving its disjoint **cycle decomposition**:

$$(a, \sigma(a), \ldots, \sigma^{r-1}(a))(b, \sigma(b), \ldots, \sigma^{s-1}(b))\ldots,$$

where it is understood that $\sigma^r(a) = a$, $\sigma^s(b) = b$, etc. Notice that r need not equal s. There need not be any relation between the lengths of different orbits for a given permutation σ.

For example, the permutation of $\{r, s, t, u\}$ defined by

$$\sigma = \begin{pmatrix} r & s & t & u \\ s & u & t & r \end{pmatrix}$$

is encoded more economically as $\sigma = (s, u, r)(t)$. Even better, we may omit all σ-orbits of length 1 and write $\sigma = (s, u, r)$. It is then understood that if the symbol t is omitted, then $\sigma(t) = t$. This is **Cauchy cycle notation** for σ.

For the identity permutation, this rule would say to omit everything, which would just leave a blank space. Instead we usually put down one "token" cycle. Often our set contains the symbol 1, in which case we usually write the identity permutation I as $I = (1)$.

It is helpful to practice the "calculus of permutations," and the exercises that follow will give you an opportunity to do this. Here are a few useful observations:

1. If $\sigma = (a, b, c, d, \ldots)(l, m, n, \ldots)\ldots$, with different letters being different elements and the cycles being disjoint, then σ^2 is gotten by taking every other letter; i.e., $\sigma^2 = (a, c, \ldots)(l, n, \ldots)\ldots$. For example, if $\sigma = (r, s, t, u)(w, x, y)$, then $\sigma^2 = (r, t)(s, u)(w, y, x)$. Likewise for higher powers. For instance taking the same σ, $\sigma^3 = (r, u, t, s)$ and $\sigma^4 = (w, x, y)$.
2. Inverses of permutations are obtained by writing each cycle backward. Thus if $\sigma = (r, s, t, u)(w, x, y)$, then $\sigma^{-1} = (u, t, s, r)(y, x, w)$.

Of course it doesn't matter which letter starts a cycle. Thus in the previous example, if we wanted to start the cycles of σ^{-1} with the same letters r and w that start the cycles of σ, this is easy: $\sigma^{-1} = (r, u, t, s)(w, y, x)$.

Also disjoint cycles may be listed in any order. Thus $\sigma^{-1} = (w, y, x)(r, u, t, s)$.

Definition. The **cycle structure** of a permutation σ is the list of cycle lengths greater than 1 for σ in nonincreasing order. Thus in the preceding example, both σ and σ^{-1} have cycle structure (4, 3), while σ^2 has cycle structure (3, 2, 2), σ^3 has cycle structure (4), and σ^4 has cycle structure (3). We'll say that the identity permutation I has cycle structure (1).

We extend the orbit language as follows.

Definition. Let H be a group of permutations of the set $S = \{a, b, c, \ldots\}$. The H-**orbit** on S containing a is the subset

$$a^H = \{h(a) : h \in H\}.$$

Thus the σ-orbit on S containing a is the $\langle\sigma\rangle$-orbit on S containing a. Here is a noncyclic example. Let R be the reflection of the complex plane across the real axis and let ρ^3 be the 180° rotation as before. Let $H = \{I, R, \rho^3, R \circ \rho^3\}$. Then H is a group of permutations of the set V of vertices of the regular hexagon and the H-orbits on V are

$$1^H = \{1, 4\} \text{ and } 2^H = \{2, 3, 5, 6\}.$$

Finally note that if $f = f(r_1, r_2, \ldots, r_n)$ is a polynomial, then the set, which Lagrange called the **set of values** of f, is exactly the S_n-orbit containing f, f^{S_n}.

You already have practiced with S_3. Let's practice our permutation calculus skills on the next larger symmetric group.

Exercises

8.1. List all of the permutations in S_4, organized by cycle structure. Record how many permutations have each given cycle structure and what the order of each permutation is. Also note when one cycle structure (other than (1)) is a "power" of another, in the sense that some permutation σ with that cycle structure is a power of some permutation τ with the other cycle structure. Do the order and power map depend on the choice of σ or just on the cycle structure?

8.2. List all the cyclic subgroups of S_4 of cardinality greater than 2.

8.3. Let Q be the square centered at $(0, 0)$ with vertices $(1, 0)$, $(0, 1)$, $(-1, 0)$ and $(0, -1)$. Number the vertices $1 = (1, 0), 2 = (0, 1), 3 = (-1, 0), 4 = (0, -1)$. Then every rotation or reflection of the square Q determines a permutation of the four vertices and so determines a permutation in S_4. List all the rotations and reflections of the square (including the identity map, which leaves every point fixed) and its corresponding permutation. This is the symmetry group D_4 we have discussed before. We may regard it as a subgroup of S_4. Describe the R-orbits on $V = \{1, 2, 3, 4\}$. Describe the ρ^2-orbits on V. Verify that $T = \{I, R, \rho^2, R \circ \rho^2\}$ is a subgroup of D_4. Write the multiplication table for the group T. Describe the T-orbits on V.

8.4. Give examples of subgroups of S_4 containing exactly 1, 2, 3, 4, 6, 8, 12, and 24 permutations. (Recall the alternating group.) What about subgroups containing 5 or 7 or ... permutations?

8.5. By looking inside D_4, show that S_4 contains nonisomorphic subgroups of cardinality 4: cyclic subgroups and subgroups that are not cyclic. Find two noncyclic subgroups of cardinality 4 inside D_4.

8.6. In the spirit of Lagrange, think of S_4 as permuting the polynomial functions in the four variables r_1, r_2, r_3, r_4. Let

$$F(r_1, r_2, r_3, r_4) = (r_1 - r_2)(r_3 - r_4)$$

and

$$G(r_1, r_2, r_3, r_4) = (r_1 + r_2)(r_3 + r_4).$$

8.6a. Find $(S_4)_F$, the set of all permutations in S_4 fixing F. Call this subgroup V_4. How many different values does F take under permutation of the variables; i.e., what is the cardinality of the S_4-orbit containing F? Also find $(S_4)_G$ and the S_4-orbit containing G.

8.6b. Find examples of functions that take exactly 1, 2, 3, 4, 6, 8, 12, and 24 values under permutation of the variables. For each such function f, write down the group $(S_4)_f$ of all permutations fixing f. Do you notice any relationship between the numbers $|f^{S_4}|$ and $|(S_4)_f|$?

8.7. In the spirit of Lagrange, consider solving the general quartic equation

$$p(x) = x^4 + cx^2 + dx + e = 0.$$

Let r_1, r_2, r_3, r_4 be the roots of p. Let $G = G(r_1, r_2, r_3, r_4)$ be as in Exercise 8.6. Explain why G is a root of a cubic equation whose coefficients are polynomials in c, d, and e. Don't actually try to find the cubic. What are the other roots of this cubic? (Call them G' and G''.) Explain why

$$r_1 + r_2 = -(r_3 + r_4).$$

Using this, "solve" for $r_1 + r_2$ in terms of G. Similarly solve for $r_1 + r_3$ in terms of G' and for $r_1 + r_4$ in terms of G''. Finally solve for r_1, r_2, r_3, and r_4 in terms of G, G', and G''. ■

By now you have probably guessed Lagrange's Theorem and maybe even figured out its proof. We'll state it in the form that Lagrange would recognize. Later we'll state it in other forms.

Lagrange's Theorem, (Version 1). *Let $f(r_1, r_2, \ldots, r_n)$ be a polynomial in the n variables $r_1, r_2, \ldots, r_n$. Let m be the number of values taken by f under permutation of the variables; i.e., $m = |f^{S_n}|$. Then m is a divisor of $n!$. Indeed*

$$n! = m \times |(S_n)_f|.$$

Proof. Let
$$f = f_1, f_2, \ldots, f_m$$
be the m different values taken by f. For each f_i there must be at least one permutation σ_i with the property that
$$\sigma_i(f) = f_i.$$
The big question is, How many permutations map f to f_i?

Let's set
$$H = \{\tau \in S_n : \tau(f) = f\};$$
i.e., $H = (S_n)_f$ is the set of all permutations fixing f.

We need two fundamental facts:

Fact 1. *Let*
$$\sigma_i \circ H = \{\sigma_i \circ \tau : \tau \in H\}.$$
Then
$$\sigma_i \circ H = \{\sigma \in S_n : \sigma(f) = f_i\}.$$

Proof. To prove that these two sets are equal, we need to show that each one is a subset of the other. First we show that the left-hand set is contained in the right-hand one. Let $\sigma = \sigma_i \circ \tau$ for some $\tau \in H$. Then
$$\sigma(f) = \sigma_i \circ \tau(f) = \sigma_i(f) = f_i.$$
Thus
$$\sigma_i \circ H \subseteq \{\sigma \in S_n : \sigma(f) = f_i\}.$$

Next we show that the right-hand set is contained in the left-hand one. Let $\sigma \in S_n$ with $\sigma(f) = f_i$. We need to concoct a permutation $\tau \in H$ such that
$$\sigma = \sigma_i \circ \tau.$$
Since permutations are invertible functions, there is only one possible choice for τ:
$$\tau = \sigma_i^{-1} \circ \sigma.$$
We must check that $\tau \in H$, i.e., that $\tau(f) = f$:
$$\tau(f) = \sigma_i^{-1} \circ \sigma(f) = \sigma_i^{-1}(f_i) = f.$$
Hence
$$\{\sigma \in S_n : \sigma(f) = f_i\} \subseteq \sigma_i \circ H.$$
So each set is contained in the other and hence they are equal, as claimed. ∎

Fact 2.
$$|\sigma_i \circ H| = |H| \text{ for all } i.$$

Proof. Define the function $\Phi : H \to \sigma_i \circ H$ by

$$\Phi(\tau) = \sigma_i \circ \tau \text{ for all } \tau \in H.$$

By definition of $\sigma_i \circ H$, this is a surjective map. So it remains to prove that it is injective. Suppose that τ and ρ are in H with $\Phi(\tau) = \Phi(\rho)$. Then

$$\sigma_i \circ \tau = \sigma_i \circ \rho$$

by definition of Φ. But then $\tau = \rho$ by the Cancellation Law in groups. Thus Φ is a bijective map and so the two sets have the same cardinality, as claimed.

Now we are done. Each permutation σ of S_n maps f to one and only one of the values f_i, so each permutation σ of S_n is contained in exactly one of the sets $\sigma_i \circ H$. So we may count the number of permutations in S_n by counting the permutations in each of these subsets and adding them up:

$$n! = |S_n| = |\sigma_1 \circ H| + |\sigma_2 \circ H| + \cdots + |\sigma_m \circ H|.$$

But each of those subsets has the *same* cardinality as H. So

$$n! = |S_n| = m \times |H| = m \times |(S_n)_f|,$$

as claimed. In particular, m is a divisor of $n!$. ∎

It is interesting that for Lagrange the significant statement was that m divides $n!$. Subsequently the other statement that $|(S_n)_f|$ divides $n!$ loomed larger. What was so special about $(S_n)_f$? After all, there are *sets* of permutations of n variables having every possible cardinality from 1 to $n!$. But if the set was of the form $(S_n)_f$ for some f, then suddenly its cardinality had to *divide $n!$*, which is much more restrictive. It must be that the fact that the set $(S_n)_f$ is *closed under multiplication* is a very powerful fact.

In fact we can go even farther. It wasn't significant that the group of permutations was S_n or that the objects being permuted were functions of n variables. Let's introduce one more term.

Definition. Let G be a group of permutations of the set $S = \{a, b, c, \dots\}$. Then

$$G_a = \{g \in G : g(a) = a\}.$$

We call G_a the **stabilizer** of a in G. (This is a small extension of our previous notation $(S_n)_f$ for the stabilizer in S_n of the function f.)

Now the proof of Lagrange's Theorem (Version 1) with obvious changes of notation proves the following theorem:

Lagrange's Orbit-Stabilizer Theorem. *Let G be a finite group of permutations (i.e. $|G| < \infty$) of the set $S = \{a, b, c, \dots\}$. Then*

$$|G| = |G_a| \times |a^G|.$$

As these realizations dawned in the minds of Ruffini, Abbati, and Cauchy, the concept of a *group* began to crystallize out of the mist of permutations.

Exercises

8.8. Let G be a group of permutations of the set $S = \{a, b, c, \dots\}$. Prove that for any $a \in S$, G_a is a subgroup of G.

8.9. Let H be a subgroup of the group of permutations G. Let σ be any permutation in G. Prove: $|\sigma \circ H| = |H|$.

(*Note*: Neither $|G|$ nor $|H|$ need be finite in Exercises 8.8 and 8.9.)

8.10. Prove Lagrange's Orbit-Stabilizer Theorem. (Imitate the proof of Lagrange's Theorem, Version 1.)

8.11. Consider the group $\Sigma(H)$ of all symmetries (rotations and reflections) of the regular hexagon H, illustrated at the beginning of this chapter. Let $V = \{I, R, \rho^3, R \circ \rho^3\}$ be the subgroup of $\Sigma(H)$ analogous to the subgroup of $\Sigma(Q)$ discussed in Exercise 3.6. Verify the orbit-stabilizer theorem for the orbits 1^V and 2^V. Notice that the numbers vary for different orbits of V.

8.12. Think of the group D_4 as a group of permutations of $\{r_1, r_2, r_3, r_4\}$ and the polynomials in those variables. Verify the orbit stabilizer theorem for the orbits $(r_1)^{D_4}$, $(r_1 + r_2)^{D_4}$, and $(r_1 r_2^2)^{D_4}$. Write down the stabilizers explicitly in each case. (Note that the set S is infinite in this case.) ∎

CHAPTER 9

ORBITS AND CAUCHY'S FORMULA

In order to get a better feeling for orbits under group actions, we digress somewhat from our historical development to prove a counting formula discovered originally by Cauchy around 1844 and rediscovered 50 years later by Frobenius. Often it is called Burnside's Counting Formula for no good reason except that Burnside included it in his textbook *Theory of Groups of Finite Order*, published in 1897.

The idea of Cauchy's formula is to count the number of orbits of a group G acting on a set S. This may seem like an arcane and uninteresting thing to do, but in fact a lot of natural things to count can be described as the set of all orbits of some finite group G acting on some other set. What's more, not only can they be described this way but this is a great boon in trying to count them. You can think of it as exploiting the natural symmetries of a situation to simplify the counting process. Chemists use it, for instance, to count the number of possible organic molecules of a given size.

Cauchy's Counting Formula. *Let G be a finite group of permutations of a finite set S. Suppose that S breaks up into n orbits under the action of G:*

$$S = \bigcup_{i=1}^{n} \mathcal{O}_i,$$

where $\mathcal{O}_i = s_i^G$. For each $g \in G$, let

$$F(g) = |\{x \in S : g(x) = x\}|.$$

Thus $F(g)$ is the number of "fixed points" of g on S. Then

$$n = \frac{1}{|G|} \sum_{g \in G} F(g);$$

i.e., the number of orbits of G on S equals the average number of fixed points.

The proof depends fundamentally on Lagrange's Orbit-Stabilizer Theorem. Note in particular the following consequence of Lagrange's Theorem: Suppose that a and b are two points in the same G-orbit $\mathcal{O}_i$; i.e., $\mathcal{O}_i = a^G = b^G$. Then by Lagrange's Theorem

$$|G_a| = \frac{|G|}{|\mathcal{O}_i|} = |G_b|.$$

So $|G_a|$ depends only on the orbit that a belongs to. Also if we add up $|G_a|$ over all the points in the orbit $\mathcal{O}_i$, then we are just adding the number $|G|/|\mathcal{O}_i|$ to itself $|\mathcal{O}_i|$ times; i.e.,

$$\sum_{a \in \mathcal{O}_i} |G_a| = |G|,$$

independent of the choice of orbit $\mathcal{O}_i$. So

$$\sum_{a \in S} |G_a| = n|G|,$$

where n is the number of G-orbits on S. So we have a formula for the number of G-orbits:

$$n = \frac{1}{|G|} \sum_{a \in S} |G_a|.$$

But it's not yet Cauchy's Formula. To get Cauchy's Formula, we need to prove that

$$\sum_{a \in S} |G_a| = \sum_{g \in G} |F(g)|.$$

To see why this is true, imagine a rectangular array whose rows are labeled by the points of S and whose columns are labeled by the permutations in G. Put the number 1 in the (s, g) position of this array if and only if

$$g(s) = s.$$

Now add up all the 1's. The sum along row a is $|G_a|$. The sum down column g is $|F(g)|$. Thus the total sum is

$$\sum_{a \in S} |G_a|$$

and also

$$\sum_{g \in G} |F(g)|,$$

proving our claim, and so proving Cauchy's Formula.

Exercises

We shall consider bracelets of the following kind. A certain number n of identically shaped beads move freely on a circular metal band. The beads come in p different colors. We shall call the beads symmetrical if they look the same when you flip the bracelet over. Otherwise we shall call them asymmetrical.

9.1. Explain how to interpret the collection of all n-beaded p-colored symmetrical bracelets as the set of all orbits of some dihedral group of rotations and reflections acting on some set.

9.2. Give a similar interpretation for the asymmetrical bracelets.

9.3. Use this to count the number of symmetrical 3-color 5-beaded bracelets.

9.4. Now count the number of asymmetrical 4-color 6-beaded bracelets.

9.5. Now count the number of symmetrical 4-color 6-beaded bracelets.

9.6. If p is a prime, give a formula for the number of asymmetrical nonmonochromatic p-color n-beaded bracelets. Is it obvious that the number is always an integer? ∎

We now apply Cauchy's Formula to the study of finite groups of rotations of $\mathbf{R}^3$. We need the following fact, apparently first established independently by Euler and Rodrigues in the 18th century:

If ρ and σ are two rotations about axes passing through $(0, 0, 0)$, then $\rho \circ \sigma$ is also a rotation about an axis passing through $(0, 0, 0)$.

As a set of bonus exercises (at the end of this chapter), we invite you to give a proof of this fact in the geometrical spirit of Chapter 2. For now we give the following more linear algebraic approach. Recall from Chapter 2 that a square matrix A is **orthogonal** if $A^t A = I$.

Definition. $SO(3)$ is the set of all 3×3 real orthogonal matrices of determinant 1.

Exercises

9.7. Prove: $SO(3)$ is a group.

9.8. Let $T : \mathbf{R}^3 \to \mathbf{R}^3$ be a linear transformation given by the formula

$$T(x, y, z) = A \begin{bmatrix} x \\ y \\ z \end{bmatrix},$$

for some $A \in SO(3)$.

9.8a. Prove: T is an isometry of $\mathbf{R}^3$. (*Hint*: Review similar arguments from Chapter 2.)

9.8b. Prove: If λ is a real eigenvalue of T, then $\lambda = \pm 1$. (*Hint*: Think geometrically and use 9.8a.)

9.8c. Prove: Either $T = I$ or T has exactly one line of eigenvectors with eigenvalue 1. (*Hint*: First argue that T must have a real eigenvalue. Why? What is the degree of the characteristic polynomial of T? Now consider separately the cases where T has three real eigenvalues and where T has exactly one real eigenvalue.)

9.8d. Suppose $T \neq I$ and let v be a nonzero vector with $T(v) = v$. Let $U = \{u \in \mathbf{R}^3 : u \cdot v = 0\}$. Prove: $T(u) \in U$ for all $u \in U$.

9.8e. Prove: If $T \neq I$, there is a real number $\theta, 0 < \theta < 2\pi$, and a basis $\{v, u, w\}$ for $\mathbf{R}^3$ where v is as in 9.8d and $u, w \in U$ such that $u \cdot w = 0$ and the matrix B for

T with respect to the basis $\{v, u, w\}$ is

$$B = \begin{bmatrix} 1 & 0 & 0 \\ 0 & \cos\theta & -\sin\theta \\ 0 & \sin\theta & \cos\theta \end{bmatrix}.$$

(Thus T is a rotation about the axis v through an angle θ.) ∎

Our goal is to study the finite subgroups of $SO(3)$, the group of all rotations of $\mathbf{R}^3$ fixing the point $(0, 0, 0)$. Denote by S^2 the unit 2-sphere in $\mathbf{R}^3$; i.e.,

$$S^2 = \{(x, y, z) : x^2 + y^2 + z^2 = 1\}.$$

Since S^2 is the set of all points at distance 1 from $(0, 0, 0)$, every nonidentity rotation T that fixes $(0, 0, 0)$ rotates S^2 about some axis L, fixing exactly two points on S^2, which we call the (north and south) **poles** of the action of T on S^2.

Now fix a nonidentity finite group G of rotations of $\mathbf{R}^3$. Associated to G we have a finite set S consisting of all of the poles for the actions of all of the nonidentity elements of G on S^2. Notice that two different rotations can have the same set of poles. Indeed if T is any rotation, then all of the nonidentity powers of $T : T^2, T^3$, etc. will have exactly the same two poles as T. So we don't know exactly how big the set S is, but we can estimate:

$$2 \le |S| \le 2(|G| - 1).$$

The largest possibility would occur if no two nonidentity elements of G have the same axis of rotation.

We claim that G acts as a finite group of permutations of the set S. This is not obvious.

Lemma. *G acts as a finite group of permutations of S.*

Proof. Let $P \in S$. Then there is a nonidentity element $g \in G$ such that $g(P) = P$. Let h be any element of G. We need to show that $h(P)$ is a pole for some nonidentity element of G. The right element to choose is hgh^{-1}. First notice

$$(hgh^{-1})(h(P)) = (hg)(h^{-1}h(P)) = hg(P) = h(P).$$

So hgh^{-1} does indeed fix $h(P)$. We must also check that $hgh^{-1} \ne I$. If $hgh^{-1} = I$, then $hg = Ih = hI$ and so $g = I$, contrary to fact. So we are done. ∎

Now that we know that the finite group G acts as a group of permutations of the finite set S, we may apply Cauchy's Counting Formula to calculate n, the number of orbits of G on S. Notice that $F(g) = 2$ for all nonidentity elements $g \in G$ and $F(I) = |S|$, so

$$n = \frac{1}{|G|}(2(|G| - 1) + |S|) = 2 + \frac{|S| - 2}{|G|}.$$

Since $|S| \ge 2$, we have that $n \ge 2$ with equality if and only if $|S| = 2$. This is the special case where every nonidentity rotation in G has the same axis. So G is just a cyclic group of rotations in the plane perpendicular to the common axis of rotation.

Since $|S| \leq 2|G| - 2$, we have

$$n \leq 2 + 2 - \frac{4}{|G|} < 4.$$

So $n \leq 3$ and we have the following proposition.

Proposition. *Let G be a finite subgroup of $SO(3)$. Then either G is a cyclic group of rotations about a fixed axis L or G has exactly three orbits on the set S of all poles of G (and $|S| = |G| + 2$).*

Now we consider the main case when the number of orbits is 3. For the sake of tradition let p, q, and r be the sizes of the three stabilizers $|G_a|, |G_b|$, and $|G_c|$ corresponding to the three orbits, listed so that $p \leq q \leq r$. Thus the three orbits have lengths $|G|/p, |G|/q$ and $|G|/r$. Since S is the union of these orbits and $|S| = |G| + 2$, we have

$$\frac{|G|}{p} + \frac{|G|}{q} + \frac{|G|}{r} = |G| + 2.$$

Dividing by $|G|$ we get

$$\frac{1}{p} + \frac{1}{q} + \frac{1}{r} = 1 + \frac{2}{|G|} > 1.$$

Notice that every pole is fixed by the identity rotation and at least one nonidentity rotation. So $2 \leq p \leq q \leq r$, and yet the sum of the reciprocals of these integers is greater than 1. This is not so easy. One way is

$$p = q = 2 \text{ and } r \text{ is arbitrary.}$$

Plugging that into the formula, we get

$$1 + \frac{1}{r} = 1 + \frac{2}{|G|}$$

and so $r = |G|/2$. Since $r = |G_c|$, this means that the pole c is fixed by half of the elements of the group G. Furthermore the orbit c^G has cardinality 2. Now the antipodal point $-c$ must also be in an orbit of size 2. So either $\{c, -c\}$ is a G-orbit or one of the other orbits has size 2, in which case by Lagrange's Theorem $|G| = 4$.

Exercise

9.9. Prove: If R and S are rotations in G with $RS = SR$, then R fixes the axis of rotation of S (as a line, not necessarily pointwise). ■

Since groups of order 4 are abelian, we see from the exercise that if $|G| = 4$, then G fixes the set $\{c, -c\}$ and so in any case, this must be a G-orbit. Since G fixes this axis L, G fixes the plane Π through $(0, 0, 0)$ perpendicular to L and acts on Π as a finite group of isometries fixing the point $(0, 0, 0)$. Thus G is isomorphic to a finite subgroup of $O(2)$ and indeed it is not hard to see that G is a dihedral group.

Exercise

9.10. Explain how the reflections in the G-action on the plane Π arise as rotations of $\mathbf{R}^3$. ■

Finally we consider the most interesting case, where $q \geq 3$. If also $p \geq 3$, then

$$\frac{1}{p} + \frac{1}{q} + \frac{1}{r} \leq \frac{1}{3} + \frac{1}{3} + \frac{1}{3} = 1,$$

a contradiction. So $p = 2$. If $q > 3$, then

$$\frac{1}{p} + \frac{1}{q} + \frac{1}{r} \leq \frac{1}{2} + \frac{1}{4} + \frac{1}{4} = 1,$$

again a contradiction. So $q = 3$. Finally since

$$\frac{1}{2} + \frac{1}{3} + \frac{1}{6} = 1,$$

we see that $r \leq 5$, and so there are exactly three possibilities. Moreover in each case we can compute $|G|$ from the formula, namely

$$\frac{1}{p} + \frac{1}{q} + \frac{1}{r} = 1 + \frac{2}{|G|}.$$

Thus

If $p = 2$, $q = 3$, and $r = 3$, then $|G| = 12$.

If $p = 2$, $q = 3$, and $r = 4$, then $|G| = 24$.

If $p = 2$, $q = 3$, and $r = 5$, then $|G| = 60$.

We are ready for the main theorem.

Theorem. *Let G be a finite group of rotations of $\mathbf{R}^3$ fixing $(0, 0, 0)$. Then one of the following five possibilities holds:*

1. $G = C_n$, $n \geq 1$, is a cyclic group of rotations about a pointwise fixed axis L through $(0, 0, 0)$ and $|G| = n$; or
2. $G = D_n$, $n \geq 2$, is a dihedral group that fixes setwise a line L through $(0, 0, 0)$ and induces a dihedral group of rotations and reflections on the plane Π through $(0, 0, 0)$ perpendicular to L, and $|G| = 2n$; or
3. $G = T$ is the tetrahedral group of all rotational symmetries of the regular tetrahedron, and $|T| = 12$; or
4. $G = O$ is the octahedral group of all rotational symmetries of the regular octahedron, and $|O| = 24$; or
5. $G = I$ is the icosahedral group of all rotational symmetries of the regular icosahedron, and $|I| = 60$.

The proof is practically complete already, and we shall not give a detailed proof. What is missing is the verification that the groups of order 12, 24, and 60 arising in the last three cases are essentially unique and may be identified as the symmetry groups of the indicated Platonic solids.

A few comments: It is fairly clear that the Platonic solids can indeed be inscribed in the 2-sphere and their symmetry groups may be computed to have the indicated cardinalities. For instance, since the icosahedron is regular, it is fairly clear that the rotational symmetry group I of the icosahedron has only one orbit on the 12 vertices. Furthermore, any rotation fixing the pair of antipodal vertices v and $-v$ must rotate the five vertices adjacent to v among themselves. (These five vertices are the vertices of a regular pentagon formed by slicing the icosahedron with a plane perpendicular to the axis through v and $-v$.) Thus the orbit v^I has size 12 and the subgroup I_v fixing v is a cyclic group of order 5. So by Lagrange's Theorem

$$|I| = |v^I||I_v| = 12 \times 5 = 60.$$

An obvious question is, Wait a minute! I thought there were *five* Platonic Solids! What happened to the other two? You are right. We haven't mentioned the cube and the dodecahedron.

Let's take a look at the orbit sizes for the group O. The orbit sizes are

$$\frac{|O|}{p} = \frac{24}{2} = 12, \quad \frac{|O|}{q} = \frac{24}{3} = 8, \text{ and } \frac{|O|}{r} = \frac{24}{4} = 6.$$

The inscribed octahedron has 6 vertices lying on S^2, and these account for the orbit of size 6. If we take the orbit of size 8, then these 8 points are the vertices of a cube inscribed in the sphere S^2. Tradition dictates that the group be called the octahedral group, but it could with equal justice be called the cube group. Another way to think of these 8 points is to take the axes of rotation passing through the centers of opposite faces of the octahedron. These axes will hit the sphere in the 8 points of an orbit, and pairs of points at the ends of an axis will be the poles for the two nonidentity rotations of the equilateral triangle faces of the octahedron.

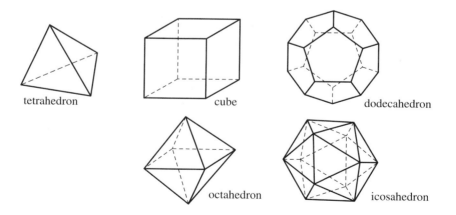

Dually, the 6 vertices of the octahedron may be thought of as the end points of the axes through the three pairs of opposite faces of the cube, and hence as the poles for the three nonidentity rotational symmetries of the square faces of the cube. Because of this dual relationship, the octahedron and the cube are called dual solids.

What about the orbit of length 12? Both the cube and the octahedron have 12 edges, and if we take a cube and an octahedron which are dual to each other, then the axes passing through the midpoints of pairs of opposite edges of the cube will also pass through the midpoints of pairs of opposite edges of the octahedron, and these will give the third orbit.

Now you can probably guess how the dodecahedron comes into the act.

Another interesting question is, Are the groups T, O, and I "new" or have we seen them before in a different guise? You can investigate this question in the following exercises.

Exercises

9.11. Give a discussion of the dodecahedron and its dual figure, similar to the one given for the octahedron and the cube.

9.12. What is the dual figure to the tetrahedron? Discuss.

9.13. The tetrahedral group T permutes the four vertices of the regular tetrahedron. Number these vertices 1, 2, 3, 4. Write each of the 12 elements of T as a permutation of $\{1, 2, 3, 4\}$ in cycle notation. What group is T?

9.14. A toy manufacturer wants to know how many differently colored tetrahedra there are, where each face is colored red, white, or blue. Can you help him? Can you see how to think of this as a problem about counting orbits? Explain. Now solve his problem for him.

9.15. The cube has four pairs of opposite vertices. Number the pairs 1, 2, 3, and 4. For each type of rotation in O (e.g., 90° rotation about an axis passing through the centers of two opposite faces; 120° rotation about an axis passing through two opposite vertices (notice that if v is a vertex of the cube, then the three vertices connected to v form the vertices of an equilateral triangle); 180° rotation about the axis passing through the midpoints of two opposite edges; etc.), write its cycle structure as a permutation of $\{1, 2, 3, 4\}$. What is your conclusion about the group O?

9.16. Now number the faces of the cube 1, 2, 3, 4, 5, 6. For each type of rotation in O, write its cycle structure as a permutation of $\{1, 2, 3, 4, 5, 6\}$.

9.17. The same toy manufacturer now wants to market red, white, and blue cubes. Using Exercise 9.16, can you help him figure out how many different red, white, and blue cubes there are? ∎

Bonus Exercises

9.18. Think about the dodecahedron inscribed in S^2. Project each edge out onto a great circle arc of S^2. Now we have a graph on the surface of S^2 with 20 vertices and 30 circular edges enclosing 12 spherical pentagons. Now imagine glueing each pair of antipodal points on S^2: (x, y, z) to $(-x, -y, -z)$. (You can't really do this in three-dimensional space, but four-dimensional people have no problem with this.) The new object is called the **real projective plane**. Our (glued) graph on the real projective plane has 10 vertices and 15 edges, since opposite vertices and opposite edges have been glued together. Draw a picture of the resulting graph. It is called **Petersen's graph**.

9.19. Give an argument that the icosahedral group I induces an action on the Petersen graph.

9.20. Form the following graph: The vertices of the graph are the unordered pairs $\{i, j\}$ where i and j are numbers between 1 and 5 with $i \neq j$. Two pairs $\{i, j\}$ and $\{k, l\}$ will be joined if

$$\{i, j\} \cap \{k, l\} = \emptyset.$$

Draw a picture of this graph. Can you see that it is the same as the Petersen graph?

9.21. Argue that the group S_5 is the full group of symmetries of the Petersen graph. Can you guess what the group I is?

9.22. A soccer ball is more or less a regular dodecahedron in which the vertices have been replaced by regular hexagons. An artistic soccer ball manufacturer wants to make soccer balls in which the hexagons are all black but the pentagons are colored either silver or gold. How many differently colored soccer balls can he manufacture? ∎

More Bonus Exercises

In this sequence of exercises you will (as promised earlier) work out a more geometrical proof that if ρ and σ are two rotations of $\mathbf{R}^3$ whose axes of rotation intersect, then $\rho \circ \sigma$ is again a rotation (fixing the point of intersection of the other two axes).

9.23. Prove: If T is an isometry of $\mathbf{R}^3$ fixing the four vertices of a tetrahedron (for instance, fixing $(0, 0, 0)$, $(1, 0, 0)$, $(0, 1, 0)$ and $(0, 0, 1)$), then T is the identity map.

9.24. Prove: If T is an isometry of $\mathbf{R}^3$ fixing $(0, 0, 0)$, then T can be written as the composition of one, two, or three reflections. T preserves orientation (has determinant 1) if and only if it can be written as the product of exactly two reflections. (*Note*: If R is any reflection, then $R \circ R = I$. So the identity map is the product of two reflections.) (*Hint*: T sends the tetrahedron with vertices $(0, 0, 0)$, $(1, 0, 0)$, $(0, 1, 0)$, and $(0, 0, 1)$ to some other tetrahedron with one vertex at $(0, 0, 0)$ and the other three vertices on the unit sphere. Now describe how to accomplish the same mapping by a sequence of one, two, or three reflections. Then use 9.23.)

9.25. Prove: If T is the product of two reflections across planes passing through $(0, 0, 0)$, then either $T = I$ or T is a rotation about an axis passing through $(0, 0, 0)$. Thus the rotations of $\mathbf{R}^3$ are precisely the isometries of $\mathbf{R}^3$ that fix $(0, 0, 0)$ and have determinant 1.

9.26. Prove: If S and T are two isometries of $\mathbf{R}^3$, both of which fix $(0, 0, 0)$ and have determinant 1, then $S \circ T$ is an isometry of $\mathbf{R}^3$ that fixes $(0, 0, 0)$ and has determinant 1. Thus the set of all rotations of $\mathbf{R}^3$ fixing $(0, 0, 0)$ forms a group. This is the group $SO(3)$. ∎

CHAPTER 9A

HAMILTON'S QUATERNIONS (OPTIONAL)

William Rowan Hamilton is best known for the Hamiltonian energy operator in classical mechanics. Hamilton was quite interested in the mechanics of three-dimensional space. In particular he was interested in an efficient method of computing rotations of $\mathbf{R}^3$. This was around 1840, when vectors and linear algebra were unknown and matrix algebra was just coming into existence. (In fact the words *vector* and *scalar* were coined by Hamilton.)

Hamilton knew that rotations in $\mathbf{R}^2$ could be computed efficiently by thinking of $\mathbf{R}^2$ as the complex numbers, in which case rotation counterclockwise through θ was simply multiplication by the complex number $e^{i\theta}$, as we saw in Chapter 6 (Exercises 6.25–27). Hamilton's idea was to define a multiplication on $\mathbf{R}^3$ that would work the same magic for rotations of 3-space.

He couldn't find one. But finally he had an inspiration. He tried looking in $\mathbf{R}^4$ instead. Still he couldn't find a "true" multiplication, but he discovered that if he was willing to give up the Commutative Law for Multiplication, then everything would work out well.

Since they were four-dimensional, Hamilton called his new objects **quaternions**. Since $\mathbf{Q}$ is already used to denote the rational numbers (quotients), it is traditional and appropriate to honor Hamilton's unique discovery by calling the quaternions $\mathbf{H}$. We shall describe three different ways of defining them.

First Definition Let

$$\mathbf{H} = \{a + bi + cj + dk : a, b, c, d \in \mathbf{R}\}.$$

Define addition in $\mathbf{H}$ "position-wise"; i.e., think of $a + bi + cj + dk$ as the "vector" (a, b, c, d) in $\mathbf{R}^4$. (This is *not* how Hamilton used the term *vector*.) Addition is ordinary vector addition in $\mathbf{R}^4$. Define multiplication by these rules:

$$i^2 = j^2 = k^2 = -1,$$

and
$$ij = k,\ jk = i,\ ki = j.$$

Extend this in the only possible way to **H** so that the Associative Law for Multiplication and the distributive laws hold and so that the real number coefficients commute with every quaternion.

Remark. *Notice that*
$$(ij)(ji) = i(jj)i = i(-1)i = (-1)i^2 = (-1)(-1) = 1.$$

But $ij = k$. So $k(ji) = 1$ and since $k^2 = -1$, multiplying on both sides by $-k$ gives $ji = -k$. Indeed we have
$$ji = -k,\ kj = -i,\ ik = -j.$$

Thus the Commutative Law for Multiplication does not hold.

Exercise

9A1. Write out the formula for the product
$$(a + bi + cj + dk)(a' + b'i + c'j + d'k).$$

9A2. Verify that if $a + bi + cj + dk$ commutes with every quaternion in **H**, then $b = c = d = 0$.

Second Definition Let
$$\mathbf{H} = \{a + v : a \in \mathbf{R},\ v \in \mathbf{R}^3\}.$$

Define addition in **H** by
$$(a + v) + (b + u) = (a + b) + (v + u),$$

where the second addition is ordinary vector addition in $\mathbf{R}^3$. If $v = 0$, call the quaternion a a **scalar**. If $a = 0$, call the quaternion v a **vector**. (This is how Hamilton used the terms *scalar* and *vector*.) Define multiplication of scalars to be ordinary multiplication in **R**. Define $av = va$ to be ordinary scalar multiplication of a vector by a scalar. Finally for vectors u and v, define
$$uv = -(u \cdot v) + (u \times v),$$

where $u \cdot v$ is the ordinary dot product in $\mathbf{R}^3$ and $u \times v$ is the cross product in $\mathbf{R}^3$. Finally extend these definitions to a general multiplication rule in **H** by the distributive law.

Remark. *The terminology of scalar and vector is Hamilton's. However, the use of dot products and cross products in $\mathbf{R}^3$ was introduced later and developed by the physicist J. Willard Gibbs. The success of this "vector algebra" and "vector calculus" in the late 19th century eclipsed Hamilton's quaternionic calculus. However, the quaternions made an important comeback with the advent of quantum mechanics.*

Exercise

9A3. Prove: If $a + u \in \mathbf{H}$, then $(a + u)^2 = -1$ if and only if $a = 0$ (i.e., $a + u = u$ is a vector quaternion) and $|u| = 1$; i.e., u lies on the unit sphere S^2 in $\mathbf{R}^3$. In particular, in $\mathbf{H}$, there are infinitely many solutions to the equation $x^2 + 1 = 0$. ∎

To motivate the next definition of the quaternions, recall that the complex numbers can be identified with 2×2 real matrices of the form

$$\begin{bmatrix} a & b \\ -b & a \end{bmatrix}$$

Third Definition Let $\mathbf{H}$ be the set of all 2×2 complex-valued matrices of the form

$$\begin{bmatrix} \alpha & \gamma \\ -\bar{\gamma} & \bar{\alpha} \end{bmatrix}$$

Addition and multiplication are ordinary matrix addition and multiplication. We identify the scalar quaternion a with the scalar matrix

$$\begin{bmatrix} a & 0 \\ 0 & a \end{bmatrix}$$

Exercises

9A4. Using the third definition for $\mathbf{H}$, prove that there is only one matrix M in $\mathbf{H}$ with $M^2 = I$, namely $-I$, the negative of the identity matrix. Thus $-I$ plays the role of -1 in this version of $\mathbf{H}$.

9A5. Since $i^2 = j^2 = k^2 = -1$, $\pm i, \pm j$, and $\pm k$ must be quaternion matrices that square to the matrix

$$-I = \begin{bmatrix} -1 & 0 \\ 0 & -1 \end{bmatrix}$$

There is a pretty obvious choice for the matrix i. Take

$$j = \begin{bmatrix} 0 & 1 \\ -1 & 0 \end{bmatrix}.$$

Find the matrix k and verify that i, j, and k satisfy the conditions in the First Definition.

9A6. Verify that quaternion addition as defined in the First and Second Definitions corresponds to matrix addition for quaternion matrices.

Recall our earlier discussion of an abstract ring as a set R with two binary operations $+$ and $\cdot$ such that $(R, +)$ is an abelian group, $\cdot$ is an associative operation and the left and right distributive laws hold. Let $M_2(\mathbf{C})$ denote the set of all 2×2 matrices with complex entries. It is a bit tedious but routine to verify that $M_2(\mathbf{C})$ forms a ring under the operations of matrix addition and multiplication. Assume that this is the case.

9A7. Verify that the Commutative Law for Multiplication does not hold in $M_2(\mathbf{C})$. Verify that $\mathbf{H}$ is a subring of $M_2(\mathbf{C})$. Make the obvious definition of $M_2(\mathbf{R})$. Clearly this is also a subring of $M_2(\mathbf{C})$. Verify that the intersection of two subrings of a ring is again a subring of that ring. To what (well-known) ring is the subring $\mathbf{H} \cap M_2(\mathbf{R})$ isomorphic?

These rings $M_2(\mathbf{C})$, $M_2(\mathbf{R})$, $\mathbf{H}$, and obvious analogues are among the most important examples of noncommutative rings.

9A8a. Verify that the set of eight matrices $\{\pm 1, \pm i, \pm j, \pm k\}$ is closed under matrix multiplication. Explain why this is enough to show that this set is a nonabelian group of order 8. It is called the **quaternion group of order 8**, Q_8.

9A8b. Prove: If Q is a group with $|Q| = 8$ having exactly one element of order 2 and six elements of order 4, then $Q \cong Q_8$. (*Hint*: Let a and b be two elements of Q of order 4 with $b \neq a^{-1}$. Argue that $\{a, a^{-1}, b, b^{-1}, ab, (ab)^{-1}\}$ are the six elements of order 4 in Q. Then argue that the map $i \to a$, $j \to b$, $k \to c$, etc., defines an isomorphism of groups.)

9A9. Define the norm and trace of a quaternion to be the determinant and trace respectively of the corresponding quaternion matrix. Verify that the quaternion norm of $a + bi + cj + dk$ is the square of the distance of (a, b, c, d) from $(0, 0, 0, 0)$ in $\mathbf{R}^4$. Verify that the quaternion matrices of trace 0 are precisely the vector quaternions. Verify that the quaternion norm on the vector quaternions corresponds to the square of the distance from $(0, 0, 0)$ in $\mathbf{R}^3$.

9A10. Verify that the non-0 quaternions form a group under multiplication. Verify that the quaternions of norm 1 form a subgroup of this group. We call this subgroup $SU(2)$. ($SU(2)$ is the 2-dimensional special unitary group. A unitary matrix A is a complex matrix such that $A\overline{A}^t = I$. A special unitary matrix is a unitary matrix of determinant 1.)

Thus the quaternions are a lot like a field of numbers. They are closed under addition, subtraction, multiplication and "division". But the multiplication does not satisfy the Commutative Law. (Thus "left" division is different from "right" division.) Mathematicians of the late 19th century were sufficiently "broad-minded" to call such things **hypercomplex numbers**. Nowadays we call $\mathbf{H}$ a **division ring**. (It has also at times been called a **skew field**.)

9A11. For Q any quaternion matrix of norm 1 and V a vector quaternion matrix, verify that QVQ^{-1} is a vector quaternion matrix of the same norm. Conclude that the mapping:

$$V \to QVQ^{-1}$$

defines a linear isometry of $\mathbf{R}^3$ thought of as the vector quaternions. Note that $Q^{-1} = \overline{Q}^t$, where $\overline{Q}$ is the matrix each of whose entries is the complex conjugate of the corresponding entry of Q. (You may use the fact that traces and determinants of similar matrices are equal.)

9A12. Let Q be the norm 1 quaternion matrix

$$\begin{bmatrix} \alpha & \gamma \\ \overline{\gamma} & \overline{\alpha} \end{bmatrix}$$

Write the matrix for the linear transformation $\rho_Q : \mathbf{R}^3 \to \mathbf{R}^3$ defined by

$$V \to QVQ{-1}$$

with respect to the basis given by the matrices i, j, and k.

9A13. Compute the determinant of the matrix for ρ_Q. Conclude that the mapping

$$Q \to \rho_Q$$

maps $SU(2)$ into $SO(3)$. ∎

With some extra work it is possible to show that this actually maps $SU(2)$ onto $SO(3)$. So Hamilton achieved his goal (more or less). Every rotation of $\mathbf{R}^3$ can be described as multiplication by quaternions of norm 1. Thus every orientation-preserving isometry S of $\mathbf{R}^3$ can be written in the form

$$S(V) = QVQ^{-1} + U$$

for some vector quaternion U and some quaternion Q in $SU(2)$.

However, the quaternion Q is not unique. There are two possible choices: $\pm Q$. It turns out that the mapping ρ is exactly a 2 to 1 "covering" of $SO(3)$ by $SU(2)$.

Exercise

9A14. Consider the subgroup $\tilde{T}$ of $SU(2)$, which maps onto the tetrahedral group T from Chapter 9 under this 2 to 1 map. What is its cardinality? Recall from 9A4 that $SU(2)$ contains only one element of order 2. Also recall the structure of the tetrahedral group T from 9.13. Using this and 9A8, argue that $\tilde{T}$ contains a quaternion subgroup of order 8 and four subgroups of order 3. $\tilde{T}$ is sometimes called the **binary tetrahedral group**. It is isomorphic to the group $SL(2, 3)$ of all 2×2 matrices of determinant 1 with entries from the field $\mathbf{Z}/3\mathbf{Z}$ of integers modulo 3, which we shall study later. ∎

Of course there is also the **binary octahedral group** $\tilde{O}$ and the **binary icosahedral group** $\tilde{I}$. The latter group is isomorphic to the group $SL(2, 5)$ of all 2×2 matrices of determinant 1 with entries from the field $\mathbf{Z}/5\mathbf{Z}$ of integers modulo 5, which we shall also study later.

This two-fold covering of $SO(3)$ by $SU(2)$ turned out much later to have deep significance with respect to the **spin** of subatomic particles. For this reason it is sometimes called the **spin covering** of $SO(3)$ by $SU(2)$. The whole subject of groups and symmetries (and broken symmetries) was to get new life and meaning in the context of 20th-century physics. We shall stick to the 19th century. However, before ending this unit, I'm sure you're all asking the question:

> If Hamilton was looking for a multiplication on $\mathbf{R}^3$, why did he reject the cross product?

I don't know the answer for sure, but I suspect that the cross product multiplication on $\mathbf{R}^3$ was simply "too different" to fit Hamilton's picture of a multiplication. Reluctantly he was willing to sacrifice the commutative law, but the cross product fails to satisfy both the commutative and the associative laws. Moreover not only is "division" not possible, there isn't even a vector that acts like 1 under cross product.

Nevertheless the cross product did not stay out of the life of "abstract algebra" for too long. The structure ($\mathbf{R}^3$, +, ×) does satisfy the distributive laws, the anticommutative law:

$$u \times v = -v \times u$$

and a "consolation prize" for the failure of the associative law called the Jacobi Identity:

$$(u \times v) \times w + (v \times w) \times u + (w \times u) \times v = 0.$$

It can be formulated in the trace 0 matrix quaternions as follows.

Exercises

9A15. Verify that if A and B are matrix quaternions of trace 0, then so are $A + B$ and $AB - BA$.

Define the **bracket product** $[A, B] = AB - BA$.

9A16. Verify that bracket multiplication satisfies the anticommutative law and the Jacobi Identity.

9A17. Verify that bracket multiplication of the quaternions i, j, and k agrees with the cross product. ∎

Definition. $su(2)$ is the **Lie algebra** of trace 0 matrix quaternions with the operations of matrix addition and bracket multiplication.

Lie algebras were introduced by Sophus Lie in the 1870s (along with Lie groups) as part of his study of the integrability of systems of differential equations. The Classification of Semisimple Lie Algebras over $\mathbf{C}$ in the 1890s by Wilhelm Killing and Élie Cartan was one of the crowning achievements of 19th-century algebra.

SECTION THREE
Numbers

CHAPTER 10 BACK TO EUCLID

CHAPTER 11 EUCLID'S LEMMA FOR POLYNOMIALS

CHAPTER 12 FERMAT AND THE REBIRTH OF NUMBER THEORY

CHAPTER 13 LAGRANGE'S THEOREM REVISITED

CHAPTER 14 RINGS AND SQUARES

CHAPTER 15 FERMAT'S LAST THEOREM (FOR POLYNOMIALS)

CHAPTER 10

BACK TO EUCLID

From the time of the Pythagoreans, the Greeks were quite interested in commensurability. What does that mean? It means: Given two line segments A and B to find a segment C such that C "measures" both A and B. In other words, A is made up of m copies of C and B is made up of n copies of C for some whole numbers m and n. Thus C may be taken as a unit of measurement, in which case A has length m units and B has length n units.

It is legendary that at some early moment, some Pythagorean made the unsettling discovery that there existed segments A and B that were not commensurable. Not only that, but these segments occurred together in the most natural of contexts. For example the diagonal of a square is not commensurable with its side, nor is the diagonal of a pentagon with its side.

Nonetheless the investigation of commensurability, like Lagrange's wonderful failed investigation of higher order quadratic formulas, led the Greeks to some beautiful mathematical discoveries.

Suppose we assume that A and B are commensurable and try to find their common measuring unit C. If A is shorter than B, then C cannot be any longer than A and we can try optimistically to measure B using A. If B is made up of exactly q copies of A for some whole number q, then we are done. If not, we have some "remainder" segment R left over and necessarily R is shorter than A.

Ah ha! Already we have a theorem. Replacing A, B, and R by their lengths α, β, and r, we have this:

Division Algorithm. *Let α, β be positive real numbers with $\alpha \leq \beta$. Then there exists a unique natural number q and a unique real number r with $0 \leq r < \alpha$, such that*

$$\beta = q\alpha + r.$$

Moreover if α and β are integers, so is r.

Although the Division Algorithm is obviously true, we shall be a bit pedantic and "prove it," based on a fundamental property of the natural numbers called the Well-Ordering Principle. Recall that **N** denotes the set of natural numbers.

The Well-Ordering Principle. *If S is any nonempty subset of **N**, then S has a least element; i.e., there exists $m \in S$ such that*

$$m \leq s \text{ for all } s \in S.$$

We shall also need a basic property of the real numbers first explicitly used by Archimedes and hence called the Archimedean Property of the Reals:

The Archimedean Property of the Reals. *Let α and β be any two positive real numbers. There exists a natural number N such that*

$$N\alpha > \beta.$$

Actually we shall use an easy variant on the Well-Ordering Principle.

Exercises

10.1. Prove (using the Well-Ordering Principle): Suppose S is a nonempty set of integers that is bounded below; i.e., assume that there exists an integer N with $N < a$ for all $a \in S$. Then there is a smallest integer in S.

10.2. Prove (using the Well-Ordering Principle): Suppose S is a nonempty set of integers that is bounded above; i.e., there exists an integer N with $a < N$ for all $a \in S$. Then there is a largest integer in S. ∎

Proof of the Division Algorithm. Let

$$Q = \{x \in \mathbf{N} : \beta - x\alpha \geq 0\}.$$

Notice that $1 \in Q$. So $Q \neq \emptyset$. By the Archimedean Property of the Reals, there exists $n \in \mathbf{N}$ with $n\alpha > \beta$. If $x \in \mathbf{N}$ with $x \geq n$, then $x\alpha \geq \alpha > \beta$ and so $\beta - x\alpha < 0$. Thus $x < n$ for all $q \in Q$; i.e., Q is bounded above by n. Hence by Exercise 10.2, there is a largest integer q in Q.

We need to show that $r = \beta - q\alpha < \alpha$. Since $q + 1 \notin Q$, $\beta - (q + 1)\alpha < 0$. Thus

$$\beta - (q + 1)\alpha = (\beta - q\alpha) - \alpha < 0,$$

and so $r = \beta - q\alpha < \alpha$, as desired.

Now we establish uniqueness. Thus suppose that $q' \in \mathbf{N}$ and r' is a nonnegative real number with $r' < \alpha$ such that

$$\beta = q'\alpha + r'.$$

Without loss, we may assume that $q' \leq q$. Then

$$q\alpha + r = q'\alpha + r',$$

and so

$$0 \leq (q - q')\alpha = r' - r.$$

As r' and r both lie on the interval $[0, \alpha)$, we have
$$0 \leq r' - r < \alpha.$$
On the other hand, if $q \neq q'$, then $q - q' \geq 1$ and so $(q - q')\alpha \geq \alpha$, a contradiction. Hence $q = q'$ and $r = r'$, as claimed. ∎

Definition. Let β and α be real numbers. If there is an integer q such that $\beta = q\alpha$, then we say that α is a **divisor** or **factor** of β, and we say that β is a **multiple** of α. (We shall mostly use these terms only when β and α are integers.)

Returning to the problem of commensurability, we arrive at this question: Suppose that when we divide β by α, we are left with a nonzero remainder r: What next? We still haven't found our common unit of measurement γ. The key observation is this: Suppose γ really exists with
$$\alpha = m\gamma \text{ and } \beta = n\gamma.$$
Then
$$r = \beta - q\alpha = n\gamma - q(m\gamma) = (n - qm)\gamma,$$
and so r is also measured by γ. Since $r < \alpha$, we may replace β and α by α and r, and try again.

There is a lovely geometric picture of this process. We can think of it as subdividing rectangles into equal squares with a remainder rectangle left over.

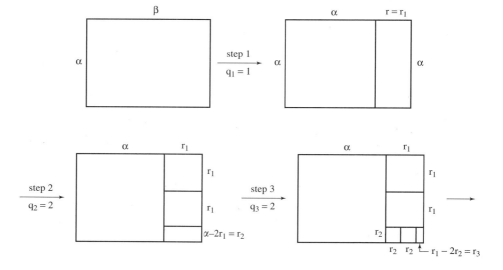

This procedure appears in Euclid's *Elements* and is called the Euclidean Algorithm. If it terminates, i.e., if you eventually reach a rectangle that subdivides exactly into squares of side equal to the short side of the rectangle, then you have found a "unit" γ (the length of the short side) that measures the original α and β (as well as all of the other segments constructed along the way).

There is an obvious way that the Euclidean Algorithm can *fail* to terminate. Suppose that the rectangle with sides α and r is similar to the rectangle with sides β and α. Then the process will continue forever producing smaller and smaller similar copies of the original rectangle. Feigning cheer in the face of adversity, the Greeks dubbed this rectangle a "Golden Rectangle" and the ratio of large to small sides the "Golden Ratio."

Exercises

10.3. Compute the Golden Ratio.

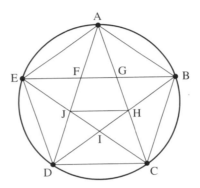

10.4. Consider the preceding picture of the regular pentagon ABCDE inscribed in a circle.

10.4a. Using the fact that equal arcs are subtended by equal angles, determine the measure of all angles in triangles ABE, BCE, and CJH.

10.4b. Prove that pentagon $FGHIJ$ is regular. Label the segment lengths $|AC| = \Delta$, $|AB| = S$, $|FG| = s$, and $|JH| = d$. Deduce these relations:

$$\frac{\Delta}{S} = \frac{d}{s},$$
$$\Delta = 2d + s,$$
$$S = d + s.$$

10.4c. Prove: If Δ and S are integers, then so are d and s.

10.4d. Prove: Δ and S cannot both be integers.

10.4e. Prove: d/s is an irrational number.

10.4f. Using the relations from (b), find an equation satisfied by the number d/s.

10.4g. Deduce that d/s is the Golden Ratio.

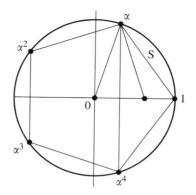

10.5. Now consider the preceding picture. Let
$$\alpha = \cos\left(\frac{2\pi}{5}\right) + i \sin\left(\frac{2\pi}{5}\right).$$

10.5a. Identify $r = \alpha + (1/\alpha)$ as the length of a line segment in the preceding picture.

10.5b. Using the Law of Sines, prove: $\Delta/S = 1/r$.

10.5c. Using 6.22, give another proof that Δ/S is the Golden Ratio.

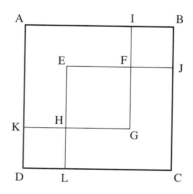

10.6. Consider the preceding picture. Let $|AB| = n$. Suppose that
$$|AI| = |JC| = m$$
$$\text{Area}(AIGK) + \text{Area}(EJCL) = \text{Area}(ABCD).$$

10.6a. Prove: $|BI| = |DL|$ and
$$\text{Area}(IBJF) + \text{Area}(KHLD) = \text{Area}(EFGH).$$

10.6b. Prove: If m and n are integers, then so are $|BI|$ and $|EF|$.

10.6c. Prove: Under the given hypotheses, m and n cannot both be integers.

10.6d. Prove (using the foregoing): $\sqrt{2}$ is an irrational number. ∎

We have now seen an instance where the Euclidean Algorithm fails to terminate. On the other hand, if we begin with two natural numbers m and n, then clearly the Euclidean Algorithm will terminate, because at each step the remainder will get smaller and yet the remainders must all be nonnegative integers. Hence after finitely many steps a remainder of 0 will be achieved and the Algorithm will have terminated. Thus if $m < n$, we will have

$$n = q_1 m + r_1,$$
$$m = q_2 r_1 + r_2,$$
$$\ldots,$$
$$r_{t-3} = q_{t-1} r_{t-2} + r_{t-1},$$
$$r_{t-2} = q_t r_{t-1} + r_t,$$
$$r_{t-1} = q_{n+1} r_t + 0.$$

As r_t divides r_{t-1} (and itself), it must also divide $r_{t-2} = q_t r_{t-1} + r_t$. Then as r_t divides r_{t-1} and r_{t-2}, it must divide r_{t-3}. Continue backward up the equations, we finally conclude that r_t is a divisor of both m and n. In modern terminology, we say that r_t is a **common divisor** of m and n. But r_t is not just any old common divisor of m and n.

Definition. Given natural numbers m and n, we say that d is the **greatest common divisor** of m and n if the following conditions hold:

(a) d is a natural number which is a divisor of both m and n; and
(b) If c is a common divisor of m and n, then c is a divisor of d.

We write $d = gcd(m, n)$ or sometimes simply $d = (m, n)$.

We remark that, using the fact about natural numbers that if c divides d, then $c \leq d$, we see that d is indeed the greatest (largest) common divisor of m and n. We also get uniqueness.

Exercises

10.7. Prove: If d and d' are two greatest common divisors of m and n, then $d = d'$.

10.8. Prove: If m and n are natural numbers and if r_t is the last nonzero remainder in the Euclidean Algorithm applied to m and n, then r_t is the greatest common divisor of m and n. ∎

The Euclidean Algorithm has an amazing corollary.

Euclid's Lemma (Version 1). *Let m and n be positive integers and let $d = gcd(m, n)$. Then d is also characterized as the least positive integer in the set*

$$\{am + bn : a, b \in \mathbf{Z}\}.$$

In particular, there exist integers a and b such that

$$d = am + bn.$$

Sketch of a Proof. Clearly since d divides m and n, d divides every number in the displayed set. So if d is in the set, it certainly must be the least positive integer in the set.

But why is d in the set at all?

This follows from the Euclidean Algorithm by "back substitution." Namely suppose again that the Euclidean Algorithm for m and n is

$$n = q_1 m + r_1,$$
$$m = q_2 r_1 + r_2,$$
$$\ldots,$$
$$r_{t-3} = q_{t-1} r_{t-2} + r_{t-1},$$
$$r_{t-2} = q_t r_{t-1} + r_t,$$
$$r_{t-1} = q_{t+1} r_t + 0.$$

Then $r_t = d$ and from the next-to-last equation

$$d = r_{t-2} - q_t r_{t-1}.$$

Using the equation before that to eliminate r_{t-1} gives

$$d = -q_t r_{t-3} + (1 + q_t r_{t-1}) r_{t-2}.$$

Continuing in this manner we eventually get $d = am + bn$ for some integers a and b, as claimed. ∎

Exercises

10.9a. Apply the Euclidean Algorithm to find $d_1 = gcd(29, 11)$.

10.9b. Solve: $29x + 11y = d_1$.

10.10a. Apply the Euclidean Algorithm to find $d_2 = gcd(294, 231)$.

10.10b. Solve $294x + 231y = d_2$.

10.11. Suppose that m and n are natural numbers with $m < n$. Suppose that n has 50 digits in its decimal expansion. Find a (good) upper bound on the number of steps in the Euclidean Algorithm for dividing n by m, where by a step we mean one application of the Division Algorithm to obtain a quotient and a remainder. ∎

Definition. A natural number p is a **prime** if $p \neq 1$ and the only natural numbers that are divisors of p are 1 and p itself.

Exercise

10.12. Suppose again that m and n are natural numbers with $m < n$ and that n has 50 digits in its decimal expansion. The Prime Number Theorem says that the number of primes less than or equal to the positive integer r is approximately $r/\ln(r)$ where $\ln(r)$ is the natural logarithm of r. Using this, estimate the worst case number of steps to find a prime divisor of n. Compare this with the estimate in 10.1. ∎

Most of you are probably familiar with the Unique Factorization Theorem for the positive integers, which asserts that any postive integer a can be written (uniquely) as a product of powers of irreducibles (primes):

$$a = p_1^{n_1} p_2^{n_2} \cdots p_r^{n_r},$$

with p_i prime and $p_1 < p_2 < \cdots < p_r$. If we also write

$$b = p_1^{m_1} p_2^{m_2} \cdots p_r^{m_r},$$

(which we may do if we allow the possibility that certain n_i's and m_j's might be 0), then most of you probably know that the gcd of a and b is

$$d = p_1^{t_1} p_2^{t_2} \cdots p_r^{t_r},$$

where $t_i = min\{m_i, n_i\}$.

However for large integers a and b, this is a very inefficient method of finding the gcd. Indeed even finding one prime factor of a can be a stupefyingly difficult task, as Exercise 10.12 suggests.

A very important consequence of Version 1 of Euclid's Lemma is the following result, which we shall call Euclid's Lemma (Version 2).

Euclid's Lemma (Version 2). *Let m, n, and a be natural numbers. Suppose that a divides mn but $gcd(a, m) = 1$. Then a divides n.*

Proof. Since $gcd(a, m) = 1$, Version 1 of Euclid's Lemma asserts that there exist integers r and s with $rm + sa = 1$. Multiplying by n gives

$$rmn + san = n.$$

Since a divides mn, there is a natural number q such that $mn = qa$. Thus

$$n = r(mn) + sn(a) = (rq)a + (sn)a = (rq + sn)a,$$

and so a is a divisor of n, as claimed. ∎

Exercises

10.13. Prove the following characterization of prime numbers: Let $p \in \mathbf{N}$ with $p > 1$. Then p is a prime if and only if the following condition holds for all natural numbers m and n:

If p divides mn, then either p divides m or p divides n.

10.14. Use mathematical induction and Exercise 10.13 to prove:

Let p be a prime and let $a_1, a_2, \ldots, a_n$ be natural numbers. If p divides the product $a_1 a_2 \cdots a_n$, then p divides a_i for some i, $1 \leq i \leq n$. ∎

Notice that the proof of both Versions 1 and 2 of Euclid's Lemma depends only on the Euclidean Algorithm, which in turn depends only on the Division Algorithm. Later we shall define a Euclidean domain to be a commutative ring in which the Cancellation Law holds and a suitable version of the Division Algorithm is valid. We shall then see that Euclid's Lemma(s) hold in all Euclidean domains. This is one of the important unifying principles of algebra.

In more general **domains of numbers** it is conventional to use the condition displayed in Exercise 10.13 as the definition of a prime and to use an analogue of our earlier definition to define an **irreducible**. Exercise 10.13 then becomes the statement that in **N** (and more generally, in Euclidean domains) primes and irreducibles are the same.

The most important corollary of Euclid's Lemma, Version 2, is the Unique Factorization Theorem for the Positive Integers, often known as

The Fundamental Theorem of Arithmetic. *Let $n \in \mathbf{N}$ with $n > 1$. Then n can be factored as a product of primes:*

$$n = p_1 p_2 \cdots p_m,$$

with p_i prime. Moreover this factorization is (essentially) unique in the sense that if

$$n = q_1 q_2 \cdots q_r$$

with q_i prime, then $r = m$ and, after suitable permutation of indices, $p_i = q_i$ for all i. (Of course, due to the Commutative Law, we cannot expect more uniqueness than this.)

Proof. First we prove existence of the prime factorization. The proof is by induction on n beginning with $n = 2$. If $n = 2$ or if n is any prime for that matter, then $n = n$ is a prime factorization and we are done. Suppose then that a prime factorization exists for all $m \in \mathbf{N}$ with $m < n$. We may assume that n is not a prime. Then by definition, there exist $a, b \in \mathbf{N}$ with $a < n, b < n$ and $n = ab$. By induction, there exist primes $p_1, p_2, \ldots, p_k$ with

$$a = p_1 p_2 \cdots p_k,$$

and there exist primes $q_1, q_2 \ldots, q_l$ with

$$b = q_1 q_2 \cdots q_l.$$

Then

$$n = ab = p_1 p_2 \cdots p_k q_1 q_2 \cdots q_l$$

is a prime factorization of n, as desired.

Next we prove uniqueness of the prime factorization. We assume false and use the Well-Ordering Principle to find a smallest natural number n having two distinct prime factorizations. Thus assume that

$$p_1 p_2 \cdots p_m = n = q_1 q_2 \cdots q_r.$$

Since p_1 is a prime and p_1 divides the product $q_1 q_2 \cdots q_r$, Exercise 10.14 says that p_1 divides one of the q_i. Renumbering we may assume that p_1 divides q_1. Then since q_1 is a prime, $p_1 = q_1$. Let $n_0 = n/p_1$. Then $n_0 < n$ and

$$p_2 \cdots p_m = n_0 = q_2 \cdots q_r.$$

Hence by minimal choice of n, $m = r$ and, after suitable further rearrangement, $p_i = q_i$ for all i, completing the proof. ∎

We also get a good characterization of the least common multiple of two integers.

Definition. Let m and n be positive integers. Then L is the least common multiple of m and n (written $L = lcm(m, n)$) if L is a positive integer satisfying:

(a) $m \mid L$ and $n \mid L$; and
(b) If M is any integer such that $m \mid M$ and $n \mid M$, then $L \mid M$.

It is clear from the definition that L is unique, if it exists.

Theorem. *Let m and n be positive integers and let $d = gcd(m, n)$. Then $L = \frac{mn}{d}$ is the $lcm(m, n)$.*

Proof. Since d divides both m and n, we have that $L = \left(\frac{m}{d}\right)n$ is a multiple of n, and $L = \left(\frac{n}{d}\right)m$ is a multiple of m. So L is a common multiple of m and n.
   On the other hand, let M be any common multiple of m and n. Write $d = am + bn$. Then

$$Md = a(Mm) + b(Mn).$$

Since m divides M, mn divides Mn. Since n divides M, mn divides Mm. But then mn divides $aMm + bMn = Md$. Write $mn = q(Md)$. As $mn = dL$, we have

$$dL = d(qM)$$

and so by the Cancellation Law,

$$L = qM.$$

Thus L divides M and so $L = lcm(m, n)$. ∎

Now let's briefly return to Pythagoras.

Definition. We call a triple (a, b, c) of natural numbers a **primitive Pythagorean triple** if $gcd(a, b) = 1$ and

$$a^2 + b^2 = c^2.$$

Exercises

10.15. The definition of primitive Pythagorean triple seems to be asymmetrical in a, b, and c. Verify that this is not the case, i.e., that $gcd(a, b) = gcd(a, c) = gcd(b, c)$.

10.16. Prove that (a, b, c) is a primitive Pythagorean triple if
$$(a, b, c) = (r^2 - s^2, 2rs, r^2 + s^2),$$
where r and s are positive integers with $r > s$, $\gcd(r, s) = 1$, and r and s are of opposite parity (i.e., one even and one odd). Verify that if r and s are both odd, then the triple is not primitive.

10.17. Make a list of the first few primitive Pythagorean triples (a, b, c).

10.18. Prove that every Pythagorean triple has the form
$$(a, b, c) = (k(r^2 - s^2), 2krs, k(r^2 + s^2))$$
for some positive integer k, with r and s satisfying the conditions in 10.16. (4.3 and 4.4 do most of the work.) Conclude that 10.16 lists *all* primitive Pythagorean triples. Explain geometrically the relationship between the right triangle with sides (a, b, c) and the right triangle with sides (ka, kb, kc). ∎

CHAPTER 11

EUCLID'S LEMMA FOR POLYNOMIALS

The Division Algorithm for Natural Numbers is the soul of simplicity *in principle*. In *practice*, the standard implementation via the Long Division Algorithm involves some subtlety.

Suppose for example we wish to divide 5722 by 361. Let's take a careful look at the beginning of the long division process. First we need to realize that implicit in the representations 361 and 5722 is the fact that we are using base 10 representations for our numbers; i.e.,

$$361 = 3(10)^2 + 6(10) + 1$$

and

$$5722 = 5(10)^3 + 7(10)^2 + 2(10) + 2.$$

We then begin the process by asking the easier question, What is the quotient when we divide $5(10)^3$ by $3(10)^2$. Since

$$5 = 1 \times 3 + 2,$$

the answer is

$$1 \times \frac{(10)^3}{(10)^2} = 1 \times 10 = 10.$$

So we take 10 as our first guess for a quotient and check the remainder:

$$5722 = 10 \times 361 + 2112.$$

Since $2112 > 361$, we are not done. But we have "reduced" the problem to the problem of dividing 2112 by 361. Since $2112 < 5722$, we have a feeling of progress.

At the second stage, however, the standard algorithm becomes more subtle. Since $3 > 2$, rather than dividing $2(10)^3$ by $3(10)^2$, we may choose instead to approximate

113

2112 by $21(10)^2$. Then since

$$21 = 7 \times 3 + 0,$$

we may take 7 as a guess for the next quotient. Alas

$$7 \times 361 = 2527 > 2112,$$

and we must try again until we finally reach the correct quotient and remainder. Numerous variants on this procedure are possible, of course. For instance we may consistently overestimate 361 by $4(10)^2$ to avoid guessing a quotient that is too large.

In any event the important point for our purposes is that the base 10 (or any other base, for that matter) representation of numbers, when written out as

$$n = a_m(10)^m + a_{m-1}(10)^{m-1} + \cdots + a_1(10) + a_0$$

makes numbers look a lot like polynomials.

And the Long Division Algorithm for Natural Numbers suggests the possibility of a Division Algorithm for Polynomials.

There are many suggestive analogies between polynomials and integers, and this has led to a very fruitful interaction between the mathematical fields of algebraic geometry (which studies polynomials) and algebraic number theory (which studies numbers). In general, results about numbers are more subtle and elusive than the corresponding results about polynomials. However, often the results about polynomials suggest what ought to be true about integers, and sometimes with luck and cleverness mathematicians have been able to prove what they guess. In Chapter 15 we shall investigate a polynomial analogue of Fermat's Last Theorem. In this chapter we shall pursue more elementary results about polynomials that mimic the results about natural numbers presented in the preceding chapter.

If $f(x) = ax^n + bx^{n-1} + \cdots$ is a polynomial with $a \neq 0$, then we say that the degree of f is $deg(f) = n$. Since n is a nonnegative integer, this gives us a way to measure how "big" the polynomial is. The 0 polynomial is a special case. Usually we say either that it has no degree or that its degree is $-\infty$.

For later reference we note that one of the basic properties of the degree map depends critically on the fact that the coefficients satisfy the Cancellation Law, a property we tend to take for granted. We make the following formal definition.

Definition. A **domain** D is a commutative ring in which the Cancellation Law holds for all $a, b, c \in D$:

$$\text{If } ab = ac, \text{ then either } a = 0 \text{ or } b = c.$$

Exercises

In the following exercises, assume that the coefficients of all polynomials lie in some domain D, unless otherwise specified.

11.1. Let D be a commutative ring. Prove that the Cancellation Law in D is equivalent to the following condition:

If a and b are in D with $ab = 0$, then $a = 0$ or $b = 0$.

11.2. If $f(x) = a_n x^n + \cdots + a_1 x + a_0$ and $g(x) = b_m x^m + \cdots + b_1 x + b_0$, what is the coefficient of x^k in $f \cdot g$? Is multiplication of polynomials commutative?

11.3. Prove: $deg(fg) = deg(f) + deg(g)$ for $f, g \neq 0$. (This suggests why $deg(0) = \infty$ or $-\infty$.)

11.4. Prove: If $h = fg$, then $deg(f) \leq deg(h)$ for $f, g, h \neq 0$. This suggests $deg(0) = \infty$, not $-\infty$. Provide a heuristic reason to set $deg(0) = -\infty$.

11.5. Let $R = \{0, 1, 2, 3\}$. Define $+_4$ and $\times_4$ as addition and multiplication "modulo" 4. In other words, $a +_4 b$ is the remainder left when $a + b$ is divided by 4, and similarly for $a \times_4 b$.

11.5a. Verify that R is a commutative ring under the operations $+_4$ and $\times_4$. (The main things to check are the two associative laws and the distributive laws. These are trivial if one of the numbers involved is 0, and the Associative Law for Multiplication is trivial if one of the numbers is 1. This shortens the checking somewhat.)

11.5b. Verify that R is not a domain.

11.5c. Give an example of two polynomials f and g with coefficients in R for which the rule $deg(fg) = deg(f) + deg(g)$ fails, and indeed $deg(fg) < deg(f)$.

11.6. Suppose $f = g + h$ with $f, g, h \neq 0$. What is the relationship between $deg(f)$ and the degrees of g and h. Does this depend on the coefficients lying in a domain? ∎

Now suppose we would like to establish a Division Algorithm for Polynomials. In other words, given two nonzero polynomials $f(x)$ and $g(x)$, we would like to find polynomials $q(x)$ and $r(x)$ with

$$f(x) = q(x)g(x) + r(x),$$

where either $r(x) = 0$ or $deg(r) < deg(g)$.

How could we begin? Suppose

$$f(x) = ax^n + bx^{n-1} + \cdots$$

and

$$g(x) = cx^m + dx^{m-1} + \cdots,$$

with $a \neq 0 \neq c$ and $m \leq n$. We could try to imitate the Long Division Algorithm. In other words, we could "approximate" g by cx^m and "approximate" f by ax^n and divide:

$$ax^n = \left(\frac{a}{c}x^{n-m}\right)(cx^m).$$

So our first guess for $q(x)$ is $q_1(x) = (a/c)x^{n-m}$. Now if we compute, we get

$$f(x) - q_1(x)g(x) = (ax^n + bx^{n-1} + \cdots) - \left(\frac{a}{c}x^{n-m}\right)(cx^m + dx^{m-1} + \cdots)$$

$$= \left(b - \frac{ad}{c}\right)x^{n-1} + \cdots = f_1(x).$$

In general $deg(f_1) \geq deg(g)$, so we have not found our quotient and remainder. But as in the Long Division Algorithm for Natural Numbers, we have improved the situation, because $deg(f_1) < deg(f)$. And so we keep on going.

There is one cautionary note. To get q_1 we had to divide a by c. So our "coefficient domain" has to allow division by nonzero numbers. Of course this is no problem if we allow arbitrary rational or real or complex coefficients. However, we would be in serious trouble if we restricted ourselves to integer coefficients.

Definition. $\mathbf{Z}[x]$ is the set of all polynomials with integer coefficients.

Exercises

11.7a. What are all the divisors of the constant polynomial 2 in $\mathbf{Z}[x]$?

11.7b. What are all the divisors of the polynomial x in $\mathbf{Z}[x]$?

11.7c. What are all the common divisors of 2 and x in $\mathbf{Z}[x]$?

11.7d. Conclude that 2 and x have a greatest common divisor d (in fact two of them), but there are no polynomials $r(x)$ and $s(x)$ in $\mathbf{Z}[x]$ with

$$2r(x) + xs(x) = d.$$

Thus Euclid's Lemma fails in $\mathbf{Z}[x]$. ∎

Recall from Chapter 5 that a set of numbers is called a field of numbers if the set contains 1 and whenever a and b are in F, then $a + b$, $a - b$, ab, and a/b (the latter provided $b \neq 0$) are also in F; i.e., you can perform all the standard operations of arithmetic (not including root extractions) without obtaining values outside of F. To avoid the problem that arose in $\mathbf{Z}[x]$, we would like to insist that our coefficients range freely over some field of numbers. However, with later generalizations in mind, we shall actually use the following more formal definition.

Definition. A field F is a set containing distinct elements 0 and 1 that is **closed** under two binary operations $+$ and $\cdot$ and satisfies the following conditions:

1. $(F, +)$ is an abelian group with identity element 0; and
2. $(F^\times, \cdot)$ is an abelian group with identity element 1, where $F^\times = F - \{0\}$; and
3. (The Distributive Law) $a(b + c) = ab + ac$ for all $a, b, c \in F$.

Exercises

11.8. Prove that the following is an equivalent definition of a field: A field F is a domain containing an element $1 \neq 0$ such that the following hold for all $a \in F$:
(a) $1 \cdot a = a$ for all $a \in F$; and
(b) To each $a \in F^\times = F - \{0\}$, there corresponds an element $a^{-1} \in F$ such that $a \cdot a^{-1} = 1$.

11.9. Using either definition explain how to define subtraction and division and prove the appropriate closure facts.

11.10. Prove: The smallest field of numbers contained in $\mathbf{C}$ is the field of rational numbers.

11.11. Prove: The set of Gaussian rationals $\mathbf{Q}(i)$ defined by

$$\mathbf{Q}(i) = \{a + bi : a, b \in \mathbf{Q}\} \leq \mathbf{C}$$

is a field.

Definition. Let F be any field of numbers. Then $F[x]$ is the set of all polynomials with coefficients in the field F, equipped with the operations of polynomial addition and multiplication. Two nonzero polynomials are equal if and only if they have the same degree and corresponding coefficients are equal.

11.12. Sketch a proof that $F[x]$ is a domain. (The Associative Law for Multiplication is pretty painful. You don't have to prove it.) Do justify carefully that the Cancellation Law is valid.

11.13. Is $F[x]$ closed under division (by nonzero polynomials)? Prove your answer. ■

Since $F[x]$ is a domain but not a field, it shares with $\mathbf{Z}$ the property that divisibility is an interesting relation. Some polynomials divide other polynomials "evenly," while others don't. And now, as in $\mathbf{Z}$, we have a Division Algorithm. We essentially gave the proof in our earlier discussion, but let's write it down a bit more formally now.

Division Algorithm for Polynomials. *Let f and g be polynomials with coefficients in the field of numbers F and with $g \neq 0$. Then there exist polynomials q (for quotient) and r (for remainder) with either $r = 0$ or the degree of r less than the degree of g such that*

$$f(x) = q(x)g(x) + r(x).$$

Proof. First we dispose of a trivial case: $f = 0$. In this case, we may take $q = r = 0$ and clearly the desired equation holds.

Next we dispose of an almost equally trivial case: $deg(f) < deg(g)$. In this case, we may take $q = 0$ and $r = f$. Again the desired equation holds and $deg(r) = deg(f) < deg(g)$, as desired.

Thus from now on we may assume that $f \neq 0$. In particular, f has a degree that is a nonnegative integer and we shall proceed by complete mathematical induction on the degree d of f. Moreover we may assume that $g \neq 0$ and $deg(g) \leq deg(f)$. We begin our induction with the smallest possible degree, that is, 0.

If f has degree 0, then $f(x) = c$ is a constant polynomial and $g(x) = a$ is a nonzero constant polynomial. The desired equation is

$$f(x) = \frac{c}{a} g(x) + 0,$$

with $q(x) = \frac{c}{a}$ and $r(x) = 0$. Notice that $\frac{c}{a}$ is in F because F, being a field, is closed under division and $a \neq 0$.

Now suppose $d > 0$ and the result is true for all polynomials $h(x)$ of degree less than d. Write

$$f(x) = a_0 x^d + a_1 x^{d-1} + \cdots$$

and

$$g(x) = b_0 x^m + b_1 x^{m-1} + \cdots$$

with a_0 and b_0 not equal to 0. As in our informal discussion before, we begin the division process by dividing the leading term of f by the leading term of g to get

$$a_0 x^d = \frac{a_0}{b_0} x^{d-m} \cdot b_0 x^m.$$

Let $q_0(x) = a_0 b_0 x^{d-m}$. Then f and $q_0 g$ have the same degree and leading coefficient, and so

$$f(x) = q_0(x) \cdot g(x) + r_0(x)$$

where $r_0(x)$ has degree at most $d - 1$. Also notice that $\frac{a_0}{b_0}$ is in our field F and so the polynomial $q_0(x)$ has its coefficients in F. As we get the coefficients of r_0 from those of f, q_0, and g by addition, subtraction, and multiplication, r_0 too has coefficients in F.

If $r_0(x)$ has degree less that m or if $r_0(x) = 0$, we are done. If not, then we may invoke complete mathematical induction to conclude that there exist polynomials $q_1(x)$ and $r(x)$ with coefficients in F such that $r(x)$ has degree less than m or $r(x) = 0$ and

$$r_0(x) = q_1(x) \cdot g(x) + r(x).$$

Plugging this into the previous equation we get

$$f(x) = (q_0(x) + q_1(x)) \cdot g(x) + r(x),$$

and this is the desired formula with $q(x) = q_0(x) + q_1(x)$. ■

Notice that this proof actually gives an algorithm for dividing one polynomial by another, although the induction argument saves us from having to go through every step of the division process.

Since $F[x]$ is a domain in which the Division Algorithm holds, the Euclidean Algorithm holds in $F[x]$. We can make the following definition.

Definition. We say that the polynomial $g \in F[x]$ is a **divisor** or **factor** of the polynomial $f \in F[x]$ if there exists a polynomial $q \in F[x]$ such that $f = q \cdot g$; i.e., if the result of the Division Algorithm is that $r(x) = 0$.

We can then speak of a common divisor of two nonzero polynomials $f(x)$ and $g(x)$ and finally of a greatest common divisor $gcd(f(x), g(x))$, namely a common divisor that is divisible by every common divisor of $f(x)$ and $g(x)$. As in the case of the integers, the Euclidean Algorithm guarantees that any two nonzero polynomials $f(x)$ and $g(x)$

have a *gcd* $d(x)$ and even gives an algorithm for finding it. (Actually we discussed *gcd*'s for **N**, not for **Z**, in the previous chapter. However, you can easily make the obvious extension.) If m and n are nonzero integers, then they always have two *gcd*'s: d and $-d$. However, we can impose uniqueness by insisting that the positive *gcd* be chosen. Likewise in $F[x]$ we can obtain uniqueness for the greatest common divisor $d(x)$ by insisting that $d(x)$ be a **monic polynomial**; i.e., a polynomial whose leading coefficient is 1.

Uniqueness is convenient but really unimportant. Crucially important, however, is the fact that the Euclidean Algorithm for Polynomials gives us

Euclid's Lemma for Polynomials (Version 1). *Let $f(x)$ and $g(x)$ be two nonzero polynomials in $F[x]$ and let $d(x)$ be a gcd of $f(x)$ and $g(x)$. Then there exist polynomials $a(x)$ and $b(x) \in F[x]$ with*

$$a(x)f(x) + b(x)g(x) = d(x).$$

Exercise

11.14. State and prove Euclid's Lemma for Polynomials, Version 2. ∎

Again the same formal argument we use in **N** (now using induction on the degrees of the polynomials and Version 2 of Euclid's Lemma for Polynomials) gives us a Unique Factorization Theorem for Polynomials. First we need some definitions.

Definition. We say that a nonconstant polynomial $p(x) \in F[x]$ is **reducible** in $F[x]$ if there exist polynomials $g(x)$ and $h(x)$ both in $F[x]$ and both of degree less than the degree of $p(x)$ such that

$$p(x) = g(x)h(x).$$

Otherwise we say that $p(x)$ is **irreducible** in $F[x]$.

Unique Factorization Theorem for Polynomials in $F[x]$. *Let $f(x)$ be a nonconstant polynomial in $F[x]$. Then there exist irreducible polynomials $p_1(x), p_2(x), \ldots, p_m(x)$ such that*

$$f(x) = p_1(x)p_2(x) \cdots p_m(x).$$

Moreover this factorization is essentially unique, in the sense that if also

$$f(x) = q_1(x)q_2(x) \cdots q_r(x)$$

with each $q_i(x)$ an irreducible polynomial in $F[x]$, then $m = r$ and, after suitable reordering, we may find nonzero constants $c_i \in F$ such that

$$q_i(x) = c_i p_i(x) \text{ for all } i.$$

Again we remark that if $f(x)$ is a monic polynomial, then we can get a stronger uniqueness statement if we insist that all of the irreducible factors also be monic polynomials. Then the factorization is unique up to reordering, just as in the Fundamental Theorem of Arithmetic.

Let's use the Division Algorithm to get a second quick proof of Descartes' Factor Theorem.

Factor Theorem. *Suppose that a is a root of the polynomial $p(x)$ (i.e., a solution of the equation $p(x) = 0$). Then $x - a$ is a factor of the polynomial $p(x)$; i.e., there exists a polynomial $q(x)$ such that $p(x) = (x - a)q(x)$.*

Proof. By the Division Algorithm, there exist polynomials $q(x)$ and $r(x)$ such that

$$p(x) = q(x) \cdot (x - a) + r(x)$$

and $r(x)$ has degree less than 1 or $r(x) = 0$. Thus $r(x) = r$ is a constant. Substitute a for x in the displayed equation and we get

$$0 = p(a) = q(a) \cdot (a - a) + r = 0 + r = r.$$

Thus $r(x) = r = 0$, as desired. ∎

Exercises

11.15a. Give an example of a polynomial $f \in \mathbf{Q}[x]$ that is irreducible in $\mathbf{Q}[x]$ but reducible in $\mathbf{R}[x]$.

11.15b. Give an example of a polynomial $g \in \mathbf{R}[x]$, that is irreducible in $\mathbf{R}[x]$ but reducible in $\mathbf{C}[x]$.

11.15c. Give an example of a polynomial $h \in \mathbf{Q}[x]$, that is irreducible in $\mathbf{Q}[x]$ but factors as $h = fg$ with both f and g irreducible in $\mathbf{R}[x]$ but reducible in $\mathbf{C}[x]$.

11.16. Let f and g be polynomials with coefficients in a field F. Suppose there exists α in some field E containing F, such that α is a common root of f and g. (For example, you might consider $F = \mathbf{Q}$, $E = \mathbf{C}$.) Prove: There is a nonconstant polynomial h with coefficients in F that divides both f and g. (*Note*: $h \in F[x]$ not $E[x]$. It's easy to find an h in $E[x]$. You need to think harder to find $h \in F[x]$.)

11.17. Prove: If $f \in F[x]$ has a multiple root in some field E of numbers containing F— i.e., if $(x - a)^2$ is a factor of $f(x)$ in $E[x]$ for some $a \in E$—then the greatest common divisor in $F[x]$ of f and its derivative polynomial f' in $F[x]$ is a nonconstant polynomial.

11.18. Prove: If $f(x)$ is an irreducible polynomial in $F[x]$ whose roots lie in some field of numbers E containing F, then the roots of f (in E) are all distinct.

11.19. Prove: The polynomial $x^n - 1$ is divisible by the polynomial $x^m - 1$ in $F[x]$ if and only if n is divisible by m. (Try "doing" the polynomial division and see what happens.)

11.20a. Factor the polynomials $x - 1, x^2 - 1, x^3 - 1, x^4 - 1, \ldots, x^{12} - 1$ into as many factors with rational coefficients as you can. Compare with 6.21. (There will always be exactly one factor that is not a factor of any previous polynomial in the series. This factor of $x^n - 1$ is called $\Phi_n(x)$, the nth **cyclotomic polynomial**. The fact that this polynomial is **irreducible** over $\mathbf{Q}$ is not easy to show. It was first proved by Gauss. We shall prove it later.)

11.20b. Look for a pattern. Can you predict the number of irreducible factors of $x^n - 1$ with rational coefficients simply from properties of the number n? Can you predict their degrees?

11.21. Let $p(x) = x^n + a_1 x^{n-1} + \cdots + a_{n-1} x + a_n$ be a monic polynomial with integer coefficients. Prove that any rational root r of p is an integer and is a divisor of a_n. (*Hint:* Write $r = a/b$ with a and b integers and $\gcd(a, b) = 1$. Plug in and clear denominators, and apply Euclid's Lemma.)

11.22. Prove that if n is a positive integer, then $\sqrt{n}$ is a rational number if and only if $n = m^2$ for some integer m. In particular $\sqrt{2}$ is irrational.

11.23. Generalize 11.21 to give a recipe for searching for possible rational roots of a polynomial $p(x) = a_0 x^n + a_1 x^{n-1} + \cdots + a_{n-1} x + a_n$ with integer coefficients. (There is a further generalization called Eisenstein's Irreducibility Criterion, which we shall see later.) ∎

CHAPTER 12

FERMAT AND THE REBIRTH OF NUMBER THEORY

Perhaps posterity will be grateful to me for having shown that the Ancients didn't know everything.

—Pierre de Fermat

Number theory languished from the time of Diophantus (c. 200 A.D.) until the 17th century. Its revival is almost singlehandedly the work of Pierre de Fermat (1601–1665). Or perhaps more accurately credit should be shared between Fermat and Leonhard Euler (1707–1783). Fermat set the stage with a dazzling array of number-theoretic discoveries communicated in letters to friends and rivals, but hardly ever accompanied by proofs. Euler around 1750 began systematically to prove all of Fermat's claims. By the time that he, Lagrange, and others were done, one of Fermat's claims had been refuted and all the rest had been proved, except for the notorious "Last Theorem," which inspired much of the development of number theory and algebraic geometry in the 19th and 20th centuries, until finally succumbing to the efforts of Andrew Wiles in 1994. Most of the remainder of this unit will be devoted to various of Fermat's "theorems" and variations thereon. We begin with his so-called Little Theorem.

Fermat's Little Theorem. *Let p be a prime and let a be a natural number. Then p divides $a^p - a$.*

This result, at least for $a = 2$, was apparently known to the Chinese mathematicians long before Fermat. As with most of his results, no one knows Fermat's proof. Probably it relied on the Binomial Theorem, which was a hot new topic among the French mathematical crowd of the 1630s and 1640s. Centered in Paris, this included Descartes and Blaise Pascal. The web of correspondence was coordinated by Father Marin Mersenne. Fermat was a corresponding member from Toulouse. The Binomial Theorem, as known to Pascal and Fermat, was the following result.

Definition. Let $\binom{n}{k}$ denote the number of ways of choosing a k element subset from a set of n objects, where k and n are natural numbers with $k \leq n$. Also set $\binom{n}{0} = 1$ for all natural numbers n.

The Binomial Theorem. $(a+b)^n = \sum_{k=0}^{n} \binom{n}{k} a^k b^{n-k}$.

The theorem is intuitively clear, since

$$(a+b)^n = (a+b)(a+b) \cdots (a+b),$$

where there are n factors on the right. By the extended Distributive Law, the product on the right equals a sum of 2^n terms obtained by choosing either an a or a b from each factor and multiplying them. By the Commutative Law, a term of the form $a^k b^{n-k}$ arises each time one chooses an a from k of the factors and a b from the remaining $n-k$ factors, hence in $\binom{n}{k}$ ways, giving the Binomial Theorem.

We outline a more formal proof by induction in the exercises that follow. First we note the following formula for $\binom{n}{k}$:

Theorem.

$$\binom{n}{k} = \frac{n(n-1) \cdots (n-k+1)}{k!} = \frac{n!(n-k)!}{k!}.$$

Proof. The number of ways of choosing an **ordered** k-tuple from a set of size n is $n(n-1) \cdots (n-k+1)$. Then the number of rearrangements of an ordered k-tuple is $k!$, giving the result. ∎

Around 1665, Newton discovered a generalization of the Binomial Theorem valid for fractional exponents, i.e. a formula for $(x+y)^{p/q}$. This played a crucial role in his development of the calculus, specifically in the differentiation and integration of power series with terms of the form $ax^{p/q}$.

Exercises

12.1. Prove: $\binom{n+1}{k} = \binom{n}{k-1} + \binom{n}{k}$ for all $k \geq 1$. (*Hint*: Let S be a set of size n. Choose one element $*$ at random in S. Every subset of S of size k either contains $*$ or does not contain $*$.)

12.2. Use Exercise 12.1 and mathematical induction to prove the Binomial Theorem.

12.3. Prove that the following two statements are equivalent:
    (a) $\binom{n}{k}$ is divisible by n for all k with $1 \leq k < n$; and

(b) n is a prime.

(*Hint*: To show that (a) implies (b), consider $\binom{n}{p}$ for p a prime divisor of n.) ∎

Now we turn to our first proof of Fermat's Little Theorem.

Proof of Fermat's Little Theorem. We proceed by induction on a. Clearly the theorem is true for $a = 1$. For the inductive step we use the Binomial Theorem:

$$(a+1)^p = a^p + pa^{p-1} + \binom{p}{2}a^{p-2} + \cdots + pa + 1.$$

By the Binomial Theorem and 12.3, each term on the right, except for the first and the last, is divisible by p. Thus

$$(a+1)^p - (a^p + 1)$$

is divisible by p. By induction $a^p - a$ is divisible by p. So the sum

$$(a+1)^p - a^p - 1 + a^p - a = (a+1)^p - (a+1)$$

is divisible by p, as claimed. ∎

Note that this theorem really depends on p being a prime. Indeed $2^4 - 2$ is not divisible by 4, nor is $2^6 - 2$ divisible by 6.

Exercises

12.4. Have you proved Fermat's Little Theorem already? Consider 9.6. Explain.

12.5. Generalize Fermat's Little Theorem slightly by allowing a to be an arbitrary integer. Prove your generalization. ∎

As noted earlier, Fermat never revealed his proof of the Little Theorem. Around 1750, Euler came up with a lovely proof. It is easiest to describe using terminology introduced somewhat later by Gauss.

Definition. We say a is **congruent** to b modulo n and write

$$a \equiv b \pmod{n}$$

if n is a divisor of $a - b$.

Let n be a positive integer. Every integer a is congruent modulo n to exactly one of the integers $0, 1, \ldots, n-1$, namely the remainder upon division:

$$a = qn + r.$$

Exercises

12.6. Suppose that $a \equiv b \pmod{n}$ and $c \equiv d \pmod{n}$.

12.6a. Prove: $a + c \equiv b + d \pmod{n}$.

12.6b. Prove: $ac \equiv bd \pmod{n}$.

12.7. Prove: If $a_1, a_2, \ldots, a_m$ and $b_1, b_2, \ldots, b_m$ are integers with $a_i \equiv b_i \pmod{n}$ for all i, then

$$a_1 a_2 \ldots a_m \equiv b_1 b_2 \ldots b_m \pmod{n}.$$

12.8a. Prove: If $ac \equiv bc \pmod{n}$ and $gcd(c, n) = 1$, then $a \equiv b \pmod{n}$.

12.8b. Give an example that shows that 12.8a becomes false if the hypothesis $gcd(c, n) = 1$ is omitted. ∎

Here's Euler's proof of Fermat's Theorem.

Euler's Proof of Fermat's Little Theorem. First notice that if a is divisible by p, then certainly a^p is divisible by p and so also $a^p - a$ is divisible by p. Hence we may assume that a is not divisible by p. Since p is a prime, this implies that $gcd(a, p) = 1$.

Now list the integers from 1 to $p - 1$:

(12A) $\qquad\qquad 1, 2, 3, \ldots, p - 1.$

Next multiply each one by a:

(12B) $\qquad\qquad a, 2a, 3a, \ldots, (p - 1)a.$

Since $gcd(a, p) = 1$, no two of ia and ja are congruent modulo p (unless $i = j$) by 12.8a. Also ia is not congruent to 0 modulo p. For if p divides ia, then by Euclid's Lemma, either p divides i or p divides a. Since $1 \leq i \leq p - 1$, this is not the case.

As noted before, every integer is congruent to one of the integers between 0 and $p - 1$ modulo p. Hence we can think of the numbers $1, 2, \ldots, p - 1$, as labels on $p - 1$ pigeonholes. We place an integer m into the pigeonhole i if $m \equiv i \pmod{p}$. By the preceding paragraph each integer listed in (12B) goes into one of the pigeonholes and no two integers listed in (12B) go into the same pigeonhole.

In other words, the relation of congruence modulo p establishes a one-to-one correspondence between list (12A) and list (12B). But then by 12.7 the product of the integers in list (12A) must be congruent modulo p to the product of the integers in list (12B) (since multiplication of integers satisfies the Commutative Law). Thus

$$(p - 1)! \equiv a^{p-1}(p - 1)! \pmod{p}.$$

Since $gcd(p, (p - 1)!) = 1$, it follows again by 12.8a that

$$1 \equiv a^{p-1} \pmod{p},$$

and so $a \equiv a^p \pmod{p}$, as claimed. ∎

Euler discovered a beautiful generalization of the Little Fermat Theorem by asking himself the following question:

I used heavily the fact that $gcd(a, p) = 1$. But where did I use the fact that p is a prime?

His answer:

Only to cancel the $(p - 1)!$ at the end.

Well that's a big *only*, because in fact $gcd(n, (n - 1)!) = 1$ *only* if n is a prime.

Exercise

12.9. Prove: $gcd(n, (n-1)!) = 1$ if and only if n is a prime. ∎

Furthermore the obvious generalization of Fermat's Little Theorem is *false* if n is not a prime. In general even if $gcd(a, n) = 1$, we do not have $a^{n-1} \equiv 1 \pmod{n}$. For example

$$3^3 \text{ is not congruent to 1 modulo 4.}$$

and

$$2^8 \text{ is not congruent to 1 modulo 9.}$$

A lesser man than Euler would give up and go on to other things. However Euler realized that you can recover cancellation if instead of multiplying *all* the integers from 1 to $n-1$ you only multiply those integers b that are **relatively prime** to n, i.e., such that $gcd(b, n) = 1$. So now Euler defines what has come to be known as Euler's φ-function:

Definition. For each positive integer n, let $\varphi(n)$ denote the number of integers b between 1 and n that are relatively prime to n.

Exercises

12.10. Determine $\varphi(n)$ for $1 \leq n \leq 24$.

12.11. If p is a prime, find a formula for $\varphi(p^n)$ and justify your claim. (*Hint*: "List" the numbers from 1 to p^n, that are multiples of p. How many numbers are on your list?)

12.12. Prove: If p is a prime and n is any positive integer, then

$$p^n = \sum_{m=0}^{n} \varphi(p^m).$$

12.13a. Let n be an even natural number. What fraction of the numbers from 1 to n are odd? If $n = 2^m$, explain how this gives a formula for $\varphi(2^m)$.

12.13b. Now suppose that n is a multiple of 6. What fraction of the numbers from 1 to n are multiples of 3? What fraction of the even numbers from 1 to n are multiples of 3? What fraction of the odd numbers from 1 to n are *not* multiples of 3. If $n = 2^a 3^b$, does this give you a formula for $\varphi(n)$?

12.13c. Generalizing, can you guess a formula for $\varphi(n)$ for *any* natural number n? ∎

Now Euler's observation is the following: List the numbers between 1 and n that are relatively prime to n:

$$a_1 = 1, a_2, \ldots, a_{\varphi(n)}.$$

If a is an integer with $gcd(a, n) = 1$, then look at the set

$$aa_1 = a, aa_2, \ldots, aa_{\phi(n)}.$$

Exercise

12.14. Let $a, b,$ and n be integers. If $\gcd(a, n) = \gcd(b, n) = 1$, then $\gcd(ab, n) = 1$. ∎

By the exercise each of the integers in the second set is relatively prime to n and by the same argument as before no two of them are congruent modulo n. Hence as before they match up modulo n with the integers in the first set and hence so do their products; i.e.,

$$a_1 a_2 \ldots a_{\varphi(n)} \equiv aa_1 aa_2 \cdots aa_{\varphi(n)} = a^{\varphi(n)} a_1 a_2 \ldots a_{\varphi(n)} \pmod{n}.$$

Now since $\gcd(a_i, n) = 1$ for all i, we *are* permitted to cancel a_i from both sides of the congruence and finally we get Euler's Theorem.

Euler–Fermat Theorem. *Let n be a positive integer and let a be an integer with $\gcd(a, n) = 1$. Then*

$$a^{\varphi(n)} \equiv 1 \pmod{n}.$$

We can go a bit farther than Euler. His argument deals with a finite set (of numbers) that is closed under an associative and commutative multiplication satisfying the Cancellation Law—in other words a finite abelian group. Euler's "real" theorem is the following.

Euler's Theorem. *Let $(A, \cdot)$ be a finite abelian group with identity element e. Then for every $a \in A$, $a^{|A|} = e$.*

Proof. List the elements of the group A:

(12C) $\qquad\qquad\qquad a_1, a_2, \ldots, a_{|A|}.$

Now multiply each element in the list by a to get the new list:

(12D) $\qquad\qquad\qquad aa_1, aa_2, \ldots, aa_{|A|}.$

Since A is closed under multiplication, every element of list (12D) is an element of list (12C). By the Cancellation Law, no two elements of list (12D) are equal. Hence by the Pigeonhole Principle, list (12D) is simply a reordering of list (12C).

Since multiplication in A satisfies the Associative and Commutative Laws, the product of all the elements of A in any order must be the same. Hence

$$a_1 a_2 \ldots a_{|A|} = aa_1 aa_2 \ldots aa_{|A|} = a^{|A|} a_1 a_2 \ldots a_{|A|}.$$

Then by the Cancellation Law we obtain

$$e = a^{|A|},$$

as claimed. ∎

We recall the notion of the **order** of an element in a group.

Definition. Let G be a group and $g \in G$. Then the **order** of g is $|\langle g \rangle|$, where $\langle g \rangle$ is the cyclic subgroup of G generated by the element g.

Also recall that if $|\langle g \rangle| < \infty$, then
$$\langle g \rangle = \{g = g^1, g^2, \ldots, g^n = e\},$$
where the $g^i \neq g^j$ for $1 \leq i < j \leq n$. Thus

Alternate Definition Let G be a group and $g \in G$. If $g^n \neq e$ for all positive integers n, then g is said to be of **infinite order**. Otherwise the **order** n of g is the smallest positive integer n such that $g^n = e$. ∎

Theorem. *Let G be a group and let g be an element of G of finite order n. Then $g^m = e$ if and only if m is a multiple of n. In particular if G is a finite abelian group, then the order n of g is a divisor of $|G|$.*

Proof. If g has order n and $g^m = e$, write
$$m = qn + r$$
for some integers q and r with $0 \leq r < n$, by the Division Algorithm. Then
$$e = g^m = g^{qn+r} = (g^n)^q g^r = e^q g^r = e g^r = g^r.$$
Thus $g^r = e$ with $0 \leq r < n$. Since n is the smallest positive integer with $g^n = e$, it follows that $r = 0$; i.e., m is a multiple of n, as claimed.

Now if G is a finite abelian group, then by Euler's Theorem, $g^{|G|} = e$ and so $|G|$ is a multiple of n; i.e., n is a divisor of $|G|$, as claimed. ∎

This raises the question: Can we use Euler's Theorem to prove the Euler–Fermat Theorem? Is there some way to think of
$$U_n = \{a \in \mathbf{N} : 1 \leq a \leq n \text{ and } gcd(a, n) = 1\}$$
as a group? The answer is of course *yes*. We can take either of the following points of view:

VIEW 1

Define a binary operation $\cdot_n$ on U_n by the following rule: Let $a, b \in U_n$. Using the Division Algorithm, write
$$ab = qn + r$$
for some integers q and r, where $0 \leq r < n$. Of course q and r depend on a and b, and we should write $q(a, b)$ and $r(a, b)$. Now define
$$a \cdot_n b = r(a, b),$$
for all $a, b \in U_n$. By the Uniqueness Part of the Division Algorithm, this product is well defined. Moreover by Exercise 12.14, since $gcd(a, n) = 1$ and $gcd(b, n) = 1$, also $gcd(ab, n) = 1$. Hence also $gcd(ab - qn, n) = 1$ and so $a \cdot_n b \in U_n$ for all $a, b \in U_n$. Hence the binary operation $\cdot_n$ is well defined and U_n is closed under this binary operation.

Exercises

12.15a. Verify that $\cdot_n$ satisfies the Commutative Law.

12.15b. Verify that $\cdot_n$ satisfies the Associative Law.

12.16. Prove: Inverses exist in U_n; i.e., for each $a \in U_n$, there exists $b \in U_n$ with $a \cdot_n b = 1$.

12.17a. Find the inverse of 5 in U_{47}.

12.17b. Find the inverse of 17 in U_{147}. (*Note*: 147 is not a prime.)

12.17c. Find the order of the element 3 in U_{19}. ∎

Notice that our remarks and 12.15 and 12.16 prove the following theorem.

Theorem. $(U_n, \cdot_n)$ *is an abelian group with identity element* 1. *Moreover* $|U_n| = \varphi(n)$. *Hence the order of every element of U_n is a divisor of $\varphi(n)$.*

Thus the Euler–Fermat Theorem is indeed Euler's Theorem applied to the group U_n.

But we promised a second view. This is a more sophisticated view. If you find it too troubling, you may put it aside. But if you wish to master more abstract mathematics, you have to come to terms with this view.

VIEW 2

In this view multiplication remains ordinary multiplication but the meaning of equality changes. On the set **Z** of all integers, we define the equivalence relation $\equiv_n$ by

$$a \equiv_n b \text{ if } a \equiv b \pmod{n}.$$

Exercise

12.18. Prove: $\equiv_n$ is an equivalence relation on **Z**. ∎

Now we consider the set **Z**/n**Z**, whose members are the equivalence classes determined by the relation $\equiv_n$. Thus each *element* of the set **Z**/n**Z** is an infinite set of numbers. For example, if $n = 5$, then **Z**/5**Z** contains the elements

$$\{\ldots, -10, -5, 0, 5, 10, \ldots\},$$
$$\{\ldots, -9, -4, 1, 6, 11, \ldots\},$$

etc. Indeed **Z**/5**Z** has exactly five elements, each of which is an infinite equivalence class of integers.

Exercise

12.19. Write down (in a similar manner) the other three elements of **Z**/5**Z**. ∎

To be careful, we should give the elements of $\mathbf{Z}/n\mathbf{Z}$ special names like $[k]$ or $\overline{k}$ or Fred. Sometimes we shall write $[k]$. But if we promise to stay alert, we can be a little sloppy and call the elements of $\mathbf{Z}/n\mathbf{Z}$, $0, 1, \ldots, n-1$, remembering that k really means the equivalence class of all integers congruent to k modulo n.

So far so good. Now we turn $\mathbf{Z}/n\mathbf{Z}$ into a ring by defining operations of addition and multiplication by the rules

$$[a] + [b] = [a+b],$$

and

$$[a] \cdot [b] = [ab].$$

Now it is not so clear that these operations are well defined because $[a] = [a']$ for infinitely many integers a'. However we notice that it is precisely the content of Exercise 12.6 that

If $[a] = [a']$ and $[b] = [b']$, then $[a+b] = [a'+b']$ and $[ab] = [a'b']$.

An advantage of this point of view is that, having verified that addition and multiplication make sense, it is very easy to verify all of the usual laws. For example,

$$[a] + ([b] + [c]) = [a] + [b+c] = [a + (b+c)] = [(a+b) + c]$$
$$= [a+b] + [c] = ([a] + [b]) + [c],$$

verifying the Associative Law for Addition. Indeed all of the laws are trivial consequences of the corresponding laws in $\mathbf{Z}$.

Whoops. Not quite. The Associative, Commutative, and Distributive Laws all hold. Hence $\mathbf{Z}/n\mathbf{Z}$ is a commutative ring. But the Cancellation Law in $\mathbf{Z}$ does not in general pass to $\mathbf{Z}/n\mathbf{Z}$. For example, in the ring $\mathbf{Z}/4\mathbf{Z}$,

$$[2] \cdot [2] = [4] = [0].$$

Exercises

12.20. Write the addition and multiplication tables for $\mathbf{Z}/2\mathbf{Z}, \mathbf{Z}/3\mathbf{Z}, \mathbf{Z}/4\mathbf{Z}, \mathbf{Z}/5\mathbf{Z}$. In which of these rings does the Cancellation Law hold? For each of these rings and each element, decide whether the element has a multiplicative inverse and if so, list it.

12.21a. Prove: If p is a prime, then every nonzero element of $\mathbf{Z}/p\mathbf{Z}$ has a multiplicative inverse. (Hence $\mathbf{Z}/p\mathbf{Z}$ is a field.)

12.21b. Prove: If n is a composite number (i.e., $n \in \mathbf{N}, n > 1$, and n is not a prime), then the Cancellation Law fails in $\mathbf{Z}/n\mathbf{Z}$. (Hence $\mathbf{Z}/n\mathbf{Z}$ is not a domain when n is composite.)

12.22. Prove: If R is any ring with operations $+$ and $\cdot$ and with multiplicative identity element 1, then the set

$$U(R) = \{r \in R : rs = 1 = sr \text{ for some } s \in R\}$$

with the operation $\cdot$ is a group with identity element 1. $U(R)$ is called the **unit group** of the ring R.

12.23. Prove: If we identify $\mathbf{Z}/n\mathbf{Z}$ with the set $\{0, 1, \ldots, n-1\}$, then for $n \geq 2$, $U_n = U(\mathbf{Z}/n\mathbf{Z})$.

12.24. Write the multiplication tables for U_5 and for U_8. Are they isomorphic groups?

12.25. Verify for all primes p, $3 \leq p \leq 17$, that U_p is a cyclic group by explicitly finding a cyclic generator and listing all of its powers. ∎

Perhaps surprisingly, we can find a setting where "honest" complex numbers behave like arithmetic modulo n. Often modular arithmetic is called "clock arithmetic." If you number the hours on a clock slightly eccentrically from 0 to 11, then telling time is arithmetic modulo 12: $11 + 3 = 2$, for example.

OK, a clock is like the unit circle S^1 and the hours on a clock are like the vertices of a regular dodecagon (12-gon). So if n is any natural number and we take the set of vertices of the regular n-gon, i.e., the n complex roots of the polynomial $x^n - 1$, then we get the set of numbers (in polar form):

$$e^{2\pi i/n}, e^{4\pi i/n}, e^{6\pi i/n}, \ldots, e^{2n\pi i/n} = 1.$$

And multiplication works like this:

$$e^{2k\pi i/n} \cdot e^{2m\pi i/n} = e^{2(k+m)\pi i/n},$$

where the addition $k + m$ is best thought of as addition modulo n. Thus if we let $\zeta = e^{2\pi i/n}$ and write ζ^k as $\zeta^{[k]}$, then our set becomes:

$$\zeta^{[1]}, \zeta^{[2]}, \zeta^{[3]}, \ldots, \zeta^{[n-1]}, \zeta^{[n]} = 1$$

and the multiplication rule becomes:

$$\zeta^{[k]} \cdot \zeta^{[m]} = \zeta^{[k]+[m]},$$

where $[k]$ and $[m]$ are the elements of $\mathbf{Z}/n\mathbf{Z}$ and addition is the addition in that ring. So multiplication of these roots corresponds exactly to addition modulo n. With this understanding, we usually drop the brackets and simply write ζ^k.

This proves the following:

Theorem. *Let $C_n = \{\zeta, \zeta^2, \ldots, \zeta^n = 1\}$ be the group of all complex nth roots of 1 with the operation of complex multiplication. Then $(C_n, \cdot)$ and $(\mathbf{Z}/n\mathbf{Z}, +)$ are isomorphic groups.*

Definition. Let n be a positive integer. We say that the complex number ζ is a **primitive** nth root of 1 if $\zeta^n = 1$ but $\zeta^m \neq 1$ for all positive integers $m < n$.

Thus in particular $e^{2\pi i/n}$ is a primitive nth root of 1.

Exercises

Let ζ be a primitive nth root of 1.

12.26. Prove: Suppose that G is a cyclic group generated by g and g has order n. Then G is generated by g^k if and only if k is a positive integer with $\gcd(k, n) = 1$. (*Hint*: Use Euclid's Lemma to write $1 = ak + bn$ for some integers a and b. Conclude that g is a power of g^k.)

12.27. Prove: ζ^k is a primitive nth root of 1 if and only if $gcd(k, n) = 1$. (*Hint*: Apply 12.26 to the cyclic group C_n of all nth roots of 1.)

12.28. Prove: For each n, there are exactly $\varphi(n)$ primitive nth roots of 1.

12.29. Prove: $n = \Sigma_d \varphi(d)$, where the sum is taken over all divisors d of n with $1 \leq d \leq n$.

12.30. Define $\Phi_n(x)$ to be the monic polynomial whose roots are the primitive nth roots of 1 (each with multiplicity 1). Then $\Phi_n(x)$ is called the nth **cyclotomic** polynomial. (Recall the discussion in 11.15.)

12.30a. Prove: $\Phi_n(x) = x^n - 1/\Pi_d \Phi_d(x)$, where the product in the denominator ranges over all divisors d of n with $1 \leq d < n$. (*Note*: $d = n$ is not included in the product.)

12.30b. Prove that $\Phi_n(x)$ has rational coefficients and degree $\varphi(n)$.

12.30c. Looking more carefully at the Division Algorithm for Polynomials, prove: Let $f(x)$ and $g(x)$ be polynomials with integer coefficients. Suppose that g is monic (i.e., has leading coefficient 1). Prove that the quotient and remainder polynomials $q(x)$ and $r(x)$ given by the Division Algorithm have integer coefficients. Prove moreover that if both f and g are monic, then q is also monic.

12.30d. Prove: If $f(x) \in \mathbf{Z}[x]$ and $g(x) \in \mathbf{Z}[x]$ with $g(x)$ monic and if $f(x) = g(x)h(x)$ with $h(x) \in \mathbf{Q}[x]$, then $h(x) \in \mathbf{Z}[x]$. Moreover if both f and g are monic, then so is h.

12.30e. Prove that $\Phi_n(x)$ is a monic polynomial with integer coefficients for all n. ∎

Let p be a prime. Since $\mathbf{Z}/p\mathbf{Z}$ is a field, all of the discussion of roots and factors of polynomials in Chapter 11 applies to the ring of polynomials with coefficients in $\mathbf{Z}/p\mathbf{Z}$. In particular, we have the following corollary of the Factor Theorem.

Theorem. *If $f(x)$ is a polynomial of degree n in $\mathbf{Z}/\mathrm{p}\mathbf{Z}[x]$, then $f(x)$ has at most n roots in $\mathbf{Z}/p\mathbf{Z}$.*

Let's think about what this says concerning the polynomial:

$$x^d - 1 \in \mathbf{Z}/p\mathbf{Z},$$

where d is a natural number. Of course it says that $x^d - 1$ has at most d roots in $\mathbf{Z}/p\mathbf{Z}$. Now if a is a root of $x^d - 1$, then

$$a^d = 1 \text{ in } \mathbf{Z}/p\mathbf{Z}.$$

Equivalently

$$a^d \equiv 1 \pmod{p}.$$

Now if $a^d = 1$ in $\mathbf{Z}/p\mathbf{Z}$, then

$$a \cdot a^{d-1} = 1 \text{ in } \mathbf{Z}/p\mathbf{Z}.$$

Thus a^{d-1} is the multiplicative inverse of a in $\mathbf{Z}/p\mathbf{Z}$. In particular $a \in U_p$ and $a^d = 1$ in the group U_p. So the order m of a as an element of the group U_p is a divisor of d. Moreover by the definition of *order*, if a has order m in U_p, then a is a root of the polynomial $x^m - 1$ in $\mathbf{Z}/p\mathbf{Z}$ and *furthermore a is not a root of the polynomial* $x^r - 1$ for *any* positive integer $r < m$. Thus

a has order m in U_p if and only if a is a **primitive** root of the polynomial

$$x^m - 1 \text{ in } \mathbf{Z}/p\mathbf{Z}.$$

Theorem. *Let p be any prime number. Let $d \in \mathbf{N}$ with $1 \leq d \leq p - 1$.*

The equation $x^d - 1 = 0$ has no primitive solutions in $\mathbf{Z}/p\mathbf{Z}$ unless d is a divisor of $p - 1$. If d is a divisor of $p - 1$, then the equation has exactly $\varphi(d)$ primitive solutions.

The group U_p has exactly $\varphi(d)$ elements of order d for each divisor d of $p - 1$. In particular, U_p is a cyclic group (with $\varphi(p - 1)$ different generators).

Proof. By the preceding comments, the second conclusion is simply a reformulation of the first conclusion in the language of groups. Hence it will suffice to prove the first conclusion.

But first notice that by the Euler–Fermat Theorem, every element of U_p has order a divisor of $p - 1$. Hence $x^d - 1 = 0$ has no primitive solutions in $\mathbf{Z}/p\mathbf{Z}$ unless d is a divisor of $p - 1$. (By a primitive solution of the equation $x^d - 1 = 0$, we mean a primitive root of the polynomial $x^d - 1$.)

For each divisor d of $p - 1$, let n_d be the number of solutions of $x^d - 1 = 0$ in $\mathbf{Z}/p\mathbf{Z}$ and let z_d be the number of primitive solutions of $x^d - 1 = 0$ in $\mathbf{Z}/p\mathbf{Z}$. Suppose that $z_d \neq 0$ and let a be a primitive solution of $x^d - 1 = 0$ in U_p. Then

$$\{a, a^2, \ldots, a^d = 1\}$$

is the set of all roots of $x^d - 1 = 0$ in $\mathbf{Z}/p\mathbf{Z}$ and forms a cyclic subgroup A_d of U_p of order d. Moreover, applying 12.26 to the cyclic group A_d, we conclude that a^k is a primitive root of $x^d - 1 = 0$ if and only if $\gcd(k, d) = 1$. In other words, if $z_d \neq 0$, then $z_d = \varphi(d)$.

Now we are ready for the punch line. The *key fact* is this: By Fermat's Little Theorem, every integer in U_p is a root of the equation $x^{p-1} - 1 = 0$ in $\mathbf{Z}/p\mathbf{Z}$. Thus $n_{p-1} = p - 1$. On the other hand, every root of $x^{p-1} - 1 = 0$ is a primitive root of $x^d - 1 = 0$ for some divisor d of $p - 1$. Thus

$$p - 1 = \sum_d z_d \leq \sum_d \phi(d) = p - 1,$$

where both sums are taken over the set of all divisors of $p - 1$. The last equality is a consequence of 12.29. But then the inequality in the middle must be an equality. Then since $z_d \leq \varphi(d)$ for all d, it follows that equality must hold for each divisor d of $p - 1$, as claimed. ∎

In particular, since $\varphi(n) \geq 1$ for all natural numbers n, it follows that U_p is a cyclic group for all primes p.

Since Euler's φ-function is popping up all over, it would be useful to have an efficient way of evaluating it. You probably guessed a good formula in 12.8. Also 12.25 may be regarded as a "recursive" formula for evaluating $\varphi(n)$ but like all recursive formulas, it can be really tedious to apply. An easier approach follows from the following amazing multiplicative property of the φ-function.

Theorem. *If* $gcd(m, n) = 1$, *then* $\varphi(mn) = \varphi(m)\varphi(n)$.

This is amazing because the "obvious" proof is totally wrong. Namely if $gcd(a, m) = 1$ and $gcd(b, n) = 1$, then it is *not true* in general that $gcd(ab, mn) = 1$. So you can't take the two obvious sets and multiply them. The correct proof is easy but clever.

Proof. Consider the following $n \times m$ rectangular array of numbers:

$$\begin{array}{cccc} 1 & 2 & \ldots & m \\ m+1 & m+2 & \ldots & 2m \\ \vdots & \vdots & \ddots & \vdots \\ (n-1)m+1 & (n-1)m+2 & \ldots & nm \end{array}$$

Clearly the numbers in the column headed by i are precisely those numbers between 1 and mn that are congruent to i modulo m. Somewhat less clearly, each column is a complete set of incongruent numbers modulo n, for if

$$im + r \equiv jm + r \pmod{n},$$

then

$$im \equiv jm \bmod n$$

and so

$$i \equiv j \pmod{n},$$

the latter because $gcd(m, n) = 1$.

Now exactly $\varphi(m)$ of the columns consist of precisely all the numbers from 1 to mn that are relatively prime to m, namely the columns headed by numbers relatively prime to m. Each of these columns contains exactly $\varphi(n)$ numbers that are relatively prime to n. Hence there are precisely $\varphi(m)\varphi(n)$ numbers in the array that are relatively prime to both m and n. These are exactly the $\varphi(mn)$ numbers between 1 and mn that are relatively prime to mn. Thus

$$\varphi(mn) = \varphi(m)\varphi(n)$$

as claimed. ∎

Exercises

12.31. Prove: If $n \in \mathbf{N}$ with $n = p_1^{a_1} p_2^{a_2} \ldots p_m^{a_m}$, then

$$\varphi(n) = p_1^{a_1-1} p_2^{a_2-1} \ldots p_m^{a_m-1}(p_1 - 1)(p_2 - 1) \cdots (p_m - 1).$$

12.32. Prove: If $n \in \mathbf{N}$ with $n = p_1^{a_1} p_2^{a_2} \ldots p_m^{a_m}$, then

$$\varphi(n) = n\left(1 - \frac{1}{p_1}\right)\left(1 - \frac{1}{p_2}\right) \cdots \left(1 - \frac{1}{p_m}\right).$$

12.33. By imitating the proof that $\varphi(mn) = \varphi(m)\varphi(n)$ when $gcd(m, n) = 1$, prove the Chinese Remainder Theorem: Let m and n be relatively prime positive integers and let k and l be integers. Then between 1 and mn, there exists exactly one integer x satisfying the pair of congruences

$$x \equiv k \quad (\text{mod } m)$$

and

$$x \equiv l \quad (\text{mod } n).$$

12.34. Now give another proof of the Chinese Remainder Theorem:
Using the basic formula $am + bn = 1$ for some integers a and b (where $\gcd(m, n) = 1$), give a "constructive" proof that
  (i) There exists a multiple of n congruent to k modulo m;
  (ii) There exists a multiple of m congruent to l modulo n; and therefore
  (iii) There exists a solution x of the pair of congruences

$$x \equiv k \quad (\text{mod } m)$$

and

$$x \equiv l \quad (\text{mod } n).$$

12.35. Prove directly (and easily): Any two solutions of the pair of congruences

$$x \equiv k \quad (\text{mod } m)$$

and

$$x \equiv l \quad (\text{mod } n)$$

differ by a multiple of mn. (This completes the second proof of the Chinese Remainder Theorem.)

12.36. Now use the Chinese Remainder Theorem to give another proof of the multiplicative property of the Euler φ-function. ∎

CHAPTER 13

LAGRANGE'S THEOREM REVISITED

Euler's Theorem is a theorem about finite abelian groups with interesting applications to numbers. Let's see if we can formulate Lagrange's Theorem as a general theorem about groups. Maybe this generalization will also yield interesting dividends.

Since Lagrange's Theorem is a theorem about one number dividing another, it will only be meaningful for **finite groups**.

Definition. A group G is said to be a finite group if $|G| < \infty$.

Until late in the 19th century, it was taken for granted that all groups were finite. When the set is finite, an alternate definition of a group can be given.

Exercises

13.1. Let G be a finite set closed under an associative multiplication. Suppose that
(a) There exists $1 \in G$ with $1 \cdot g = g = g \cdot 1$ for all $g \in G$; and
(b) The Cancellation Law holds; i.e., if $g, h, k \in G$ with
$$g \cdot h = g \cdot k,$$
then $h = k$.
Prove that G is a finite group; i.e., that inverses exist.

13.2. Give an example to show that the preceding conditions on an infinite set do *not* guarantee that the set is a group. (*Hint*: "Multiplication" can be addition and 1 can be 0.) ∎

If we think about the statement of Lagrange's Theorem, especially Version 1, there doesn't seem to be much hope of turning it into a theorem about finite groups, since it's so deeply connected with functions and permutations. But if we look at the proof and think about what is important there, things begin to be more hopeful.

We see that the proof is all about a group S_n and a subgroup $H = (S_n)_f$ together with a collection of subsets, all of the form $\sigma \cdot H$ for some $\sigma \in S_n$. Such subsets play an important role in group theory. We call them **(left) cosets** of H.

Definition. Let G be a group, H a subgroup of G, and s an element of G. The subset

$$sH = \{sh : h \in H\}$$

is called a **left coset** of H in G. A similar construction yields the **right coset** Hs.

Whenever we have a group G and a subgroup H, we may define two equivalence relations on the set G:

Definition. Let G be a group and H a subgroup of G. Define a relation $_H\!\equiv$ on G by

$$x\,_H\!\equiv y \text{ if and only if } x^{-1}y \in H.$$

Also define a relation $\equiv_H$ on G by

$$x \equiv_H y \text{ if and only if } xy^{-1} \in H.$$

Exercises

Let G be a group and H a subgroup of G.

13.3a. Prove: $_H\!\equiv$ is an equivalence relation on G.

13.3b. Prove: The $_H\!\equiv$-equivalence classes are the left cosets of H in G.

13.3c. Prove: $\equiv_H$ is an equivalence relation on G and the $\equiv_H$-equivalence classes are the right cosets of H in G.

13.4. If $G = \mathbf{Z}$, $n \in \mathbf{N}$, and $H = n\mathbf{Z}$, what are the relations $_H\!\equiv$ and $\equiv_H$?

13.5. Prove: If G is an abelian group, then for every subgroup H of G, $_H\!\equiv$ and $\equiv_H$ are the same relation.

13.6. Prove: $_H\!\equiv$ and $\equiv_H$ are the same relation if and only if $gH = Hg$ for every $g \in G$. (When this condition holds, we say that H is a **normal** subgroup of G.)

13.7. Think of an example of a group G and a subgroup H for which the relations $_H\!\equiv$ and $\equiv_H$ are different.

13.8. Prove: If $t \in sH$, then $sH = tH$.

13.9. Prove: If $t \notin sH$, then $sH \cap tH = \emptyset$.

13.10. Prove: There is a one-to-one correspondence between the sets H and sH for any left coset sH of H. ∎

Now we can state and prove Lagrange's Theorem for finite groups. You have done most of the work in the exercises.

Lagrange's Theorem (Version 2). *Let G be a finite group and H a subgroup of G. Then $|H|$ divides $|G|$.*

Proof. Since G is a finite set, the equivalence relation $\equiv_H$ defines a disjoint partition of G into finitely many equivalence classes, i.e., left cosets of H:

$$G = \bigcup_{i=1}^{r} s_i H.$$

Then

$$|G| = \sum_{i=1}^{r} |s_i H|.$$

But then by 13.10, $|s_i H| = |H|$ for all i, and so

$$|G| = \sum_{i=1}^{r} |H| = r|H|. \qquad \blacksquare$$

Often we write $|G : H|$ to denote the number of left (or right) cosets of H in G, and call this number the **index** of H in G. Then Lagrange's Theorem takes the slightly sharper form

$$|G| = |H| \cdot |G : H|.$$

Compare this with the Orbit-Stabilizer Theorem:

$$|G| = |G_a| \cdot |a^G|.$$

They are very much the same result. Indeed the group G acts as a group of permutations of the set $S = \{sH : s \in G\}$ of all left cosets of H in G via left multiplication. The stabilizer G_H of the "point" H is the subgroup H and the orbit H^G of the "point" H is the entire set S. As $|S| = |G : H|$, this gives another proof of this version of Lagrange's Theorem.

Plus ça change, plus c'est la même chose, as Lagrange himself might have said. There are only a few great ideas in mathematics, but a lot of devilishly clever variations on them.

Exercises

13.11. Prove: Suppose that σ is a permutation of the set S with σ of order p where p is a prime. Then every σ-orbit on S has cardinality either 1 or p. (You may use Lagrange's Theorem or you may give a more elementary proof.)

13.12a. Prove: Suppose that σ is a permutation of the set S with σ of order p where p is a prime. Let

$$Fix(\sigma) = \{x \in S : \sigma(x) = x\}.$$

Then

$$|S| \equiv |Fix(\sigma)| \pmod{p}.$$

(*Hint*: Verify that $x \in Fix(\sigma)$ if and only if $\{x\}$ is a σ-orbit on S. Then use 13.11.)

13.12b. Show by example that there is no sensible (and true) generalization of 13.12a to the case where the order of σ is an arbitrary composite number.

13.13. If p is a prime and G is a group with $|G| = p$, then $G \cong C_p \cong \mathbf{Z}/p\mathbf{Z}$. (Recall that C_p is the cyclic group of all rotational symmetries of the regular p-gon. Also recall that we have proven earlier that any two cyclic groups of the same cardinality are isomorphic as groups. Hence it suffices to show that G is cyclic. Henceforth we shall use the notation C_n for the cyclic group of cardinality n and reserve $\mathbf{Z}/n\mathbf{Z}$ for the ring of integers modulo n.) ∎

Arthur Cayley was one of the first mathematicians to champion the idea of thinking about abstract (finite) groups, starting in the mid-19th century. He also introduced the idea of constructing a multiplication table for a group like the usual multiplication tables for numbers, in which the rows and columns are labeled by the group elements (typically with the same ordering for both rows and columns and with the element 1 labeling the first row and the first column) and the (g, h) position of the array is the product gh. Clearly this multiplication table (or Cayley table) contains complete information about the group (as abstract group) and so determines the group up to isomorphism of groups. Two groups G and H are isomorphic if and only if, upon relabeling every entry in the Cayley table for G by an element of H using a suitable one-to-one map (namely the isomorphism function $f : G \to H$), we get a correct Cayley table for H. Unfortunately the size of the Cayley table for a group of cardinality n is n^2 and so it is only practical to write down a Cayley table for very small groups. More sophisticated reasoning is appropriate for determining the isomorphism or nonisomorphism of larger groups.

The number of nonisomorphic groups of a given cardinality n tends to be correlated with the number of prime divisors of n. If n is a prime, you have shown in 13.13 that there is, up to isomorphism, only one group of cardinality n. In the next set of exercises, you will consider the next easiest case: groups of cardinality $2p$ for p a prime.

Exercises

13.14. Let G be a group in which every element has order 1 or 2.

13.14a. Prove: G is an abelian group. (*Hint*: Let $a, b \in G$ and consider $(ab)^{-1}$.)

13.14b. Prove: If $|G| > 2$, then G has a subgroup H of cardinality 4. In fact, let a and b be any two nonidentity elements of G. Construct the Cayley table for the smallest subgroup H of G containing both a and b. Verify that H is a noncyclic group of cardinality 4.

13.15. Prove: If G is a group of cardinality $2p$ with p a prime, then G contains an element of order p (and hence a cyclic subgroup of cardinality p). (*Hint*: First show that if G is cyclic, then G contains an element of order p. Next use 13.14 to analyze the case when every element of G has order 1 or 2.)

13.16. Prove: Let G be a group and H and K subgroups of G. Then $H \cap K$ is a subgroup of G.

13.17. Prove: Let G be a group and H and K subgroups of G of cardinalities m and n respectively. Suppose that $H \cap K = \{1\}$. Then $|HK| = mn$, where $HK = \{hk : h \in H, k \in K\}$. (*Note*: This is not obvious. Why not? Prove by example that the statement can be false if H is a subgroup but K is only a subset of G containing 1.)

13.18. Prove: If G is a group of cardinality $2p$ with p an odd prime, then G has exactly one subgroup H of cardinality p. (Use 13.15 and 13.17.)

13.19. Prove: If G is a group of cardinality $2p$ and G is not a cyclic group, then G has a subgroup H of order p and an element g of order 2 such that
(a) Every element of G can be written as h^i or gh^i for some i between 0 and $p - 1$; and
(b) $h^i g = gh^{-i}$ for all i.
Argue that these rules uniquely determine the multiplication in G and so G is uniquely determined up to isomorphism. Argue that G is isomorphic to the group D_p of all rotations and reflections of the regular $2p$-gon. Thus up to isomorphism, there are exactly two groups of order $2p$: C_{2p} and D_p. ∎

Remark. *It can be proved more generally that if p and q are prime numbers, then there are at most two nonisomorphic groups of cardinality pq. If $p = q$, there are always two. If $p < q$, there are two if and only if $p \mid q - 1$.*

It took some effort in 13.15 to show that if G is a group with $|G| = 2p$, then G contains an element of order p. Remarkably this is just a special case of a much more general theorem discovered by Cauchy.

Cauchy's Theorem. *Let G be a finite group. If p is a prime divisor of $|G|$, then G contains an element of order p.*

Even more remarkably there is a very elementary (but sneaky) proof of Cauchy's Theorem. (This is not Cauchy's proof, which is very interesting but quite a bit more complicated.)

Cauchy's Theorem. *Let G be a finite group such that $|G|$ is divisible by the prime p. Then G contains an element of order p.*

Proof. The trick is to consider the set
$$S = \{(g_1, g_2, \ldots, g_p) \in G \times \cdots \times G : g_1 g_2 \cdots g_p = 1\}$$
and the left-shift function $\lambda : S \to S$ defined by
$$\lambda(g_1, g_2, \ldots, g_p) = (g_2, g_3, \ldots, g_p, g_1)$$
for all $(g_1, g_2, \ldots, g_p) \in S$. First notice that the image of λ is indeed in S because if $g_1(g_2 \cdots g_p) = 1$, then $g_2 \cdots g_p = g_1^{-1}$ and so also $(g_2 \cdots g_p)g_1 = 1$. Next notice that obviously λ is an invertible map, whose inverse is the right-shift operator:
$$\rho(g_1, g_2, \ldots, g_p) = (g_p, g_1, \ldots, g_{p-1}).$$
Hence λ is a permutation of the set S.

Now we count $|S|$. The first $p-1$ entries of $(g_1, g_2, \ldots, g_p)$ may be chosen arbitrarily and then g_p is uniquely determined as $(g_1 g_2 \cdots g_{p-1})^{-1}$. So $|S| = |G|^{p-1}$. In particular $|S|$ is divisible by p.

Now we apply 13.12a to conclude that $|Fix(\lambda)|$ is also divisible by p. But clearly
$$Fix(\lambda) = \{(g, g, \ldots, g) \in G \times \cdots \times G : g^p = 1\}.$$

Thus $Fix(\lambda)$ contains $(e, e, \ldots, e)$, where e is the identity element of G, together with all $(g, g, \ldots, g)$ such that g has order p. As $|Fix(\lambda)|$ is divisible by p, there must be at least $p - 1$ elements of G of order p, completing the proof. (In fact we have shown that G contains $kp - 1$ elements of order p for some positive integer k.) ∎

Exercise

13.20. Let G be a group such that $|G|$ is divisible by the prime p. Prove that the number n_p of cyclic subgroups of G of order p has the form $n_p = kp + 1$ for some integer $k \geq 0$. (*Hint*: Use the remark at the end of the proof of Cauchy's Theorem. Also argue that the number of elements of G of order p is $n_p(p - 1)$.) ∎

In 1872, Ludwig Sylow published an important generalization of Cauchy's Theorem and 13.20. He proved

Sylow's Theorem. *Let G be a finite group such that $|G|$ is divisible by the prime power p^a but not by p^{a+1}. Then G has a subgroup P with $|P| = p^a$. Moreover the number s_p of subgroups of G of cardinality p^a has the form $s_p = kp + 1$ for some integer $k \geq 0$.*

You will have an opportunity to work out a proof of Sylow's Theorem in Exercises 19.15 and 19.16 of the Appendix to Chapter 19. The proof relies only on the idea of a group of permutations of a set and the following small extension of 13.12a:

Theorem. *Let P be a finite group of permutations of a set S such that $|P|$ is a power of a prime p. Let*

$$Fix(P) = \{x \in S : g(x) = x \text{ for all } g \in P\}.$$

Then

$$|Fix(P)| \equiv |S| \pmod{p}.$$

More Exercises

13.21. Verify that the symmetric group S_4 does not contain an element of order 6. (So Cauchy's Theorem does not extend to nonprime divisors.) Let A_4 denote the alternating subgroup of Σ_4 of cardinality 12. List all of the elements of A_4 of order 2. Verify that A_4 does not contain a subgroup of cardinality 6. (Thus the converse of Lagrange's Theorem is in general *false*.)

13.22. Prove: If G is a group with $|G| = 15$, then $G \cong C_{15}$. (*Hint*: Use 13.20 to count the number of elements of G of order 3 and 5. By way of contrast there are 14 nonisomorphic groups of cardinality 16.) ∎

CHAPTER 14

RINGS AND SQUARES

We now return to the example of the ring of **Gaussian integers**, which we first studied briefly in Chapter 6. We know from 6.25 that the Gaussian integers form a ring of numbers. Indeed since the Cancellation Law holds in **C** and every subring thereof, **Z**[i] is a domain. Also the norm map on the complex numbers,

$$N(a + bi) = a^2 + b^2,$$

when restricted to **Z**[i] takes values in the nonnegative integers and so defines a size function on **Z**[i]. So there is hope of proving a Euclidean Algorithm for **Z**[i].

Fermat determined exactly which natural numbers n have the property that

$$n = a^2 + b^2$$

for some integers a and b. Clearly this can be reformulated as the question,

> Which natural numbers n occur as the norm of a Gaussian integer?

We shall see that this reformulation gives a way to solve Fermat's problem. Specifically we shall use properties of the Gaussian integers to establish the main part of Fermat's result, namely to determine which primes are expressible as a sum of two squares.

Exercise

14.1. Using properties of the norm map, prove that the set S of nonnegative integers defined by

$$S = \{n : n = a^2 + b^2\}$$

is closed under multiplication. Give an explicit expression for $(a^2 + b^2)(c^2 + d^2)$ as a sum of two squares. ■

It is helpful (and fun) to establish a Division Algorithm for Gaussian integers.

Division Algorithm in Z[i]. *Let m and n be Gaussian integers with $m \neq 0$. There exist Gaussian integers q and r with*

$$n = qm + r$$

where $N(r) < N(m)$.

Proof. The key is to understand the multiples of m geometrically. If a is an ordinary integer, then am is a "vector" on the line through 0 and m. So the ordinary integer multiples of m are spaced out evenly at intervals of length $\sqrt{N(m)}$ along the line through the "vector" m.

Now multiplication by i, as we saw, is geometrically equivalent to rotating counterclockwise by 90°. So im is a "vector" perpendicular to m and the ordinary integer multiples of im are again spaced out evenly at intervals of length $\sqrt{N(m)}$ along the line through the "vector" im, i.e., the line through 0 perpendicular to the line through 0 and m.

Finally if we take an arbitrary Gaussian integer $a + bi$, then

$$(a + bi)m = am + b(im)$$

and so we get a "lattice" consisting of squares of side $\sqrt{N(m)}$ with the multiples of m sitting at the lattice points.

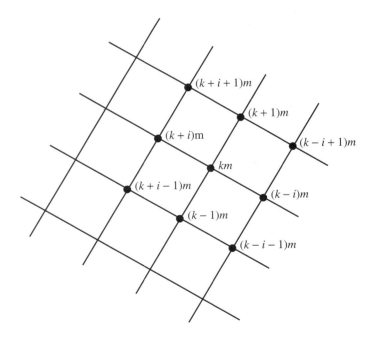

Now, if we want to divide n by m, we must look for a square in our m-lattice containing the point n. (If n lies "on" the lattice, there will be more than one choice for this square.) Let us say that the vertices of this square are km, $(k + 1)m$, $(k + i)m$ and $(k + 1 + i)m$, for some Gaussian integer k. We will choose q in the set $\{k, k + 1, k + i, k + 1 + i\}$ so as to minimize the distance from n to qm, i.e. to minimize $N(r)$ where $r = n - qm$.

What is the greatest this distance could be?

Obviously, the worst case scenario occurs when n is at the center of the square. Then

$$n = qm + \frac{1+i}{2}m.$$

So we see that for $r = n - qm$, the worst case is $r = \left(\frac{1+i}{2}\right)m$. So

$$N(r) \leq N\left(\frac{1+i}{2}m\right) = N\left(\frac{1+i}{2}\right)N(m) = \frac{1}{2}N(m) < N(m).$$

(Geometrically all we are saying is that the distance from a vertex of a square to the center of the square is $1/\sqrt{2}$ times the length of a side of the square. In particular it is shorter than the length of a side.)

This completes the proof of the Division Algorithm for $\mathbf{Z}[i]$. ∎

Exercise

14.2. Show by example that the q and r in the Division Algorithm in $\mathbf{Z}[i]$ are not in general unique. That is, find m and n nonzero in $\mathbf{Z}[i]$ such that

$$m = q_1 n + r_1 \text{ and } m = q_2 n + r_2 \text{ with } N(r_1) < N(n) \text{ and } N(r_2) < N(n)$$

but $q_1 \neq q_2$ and $r_1 \neq r_2$. ∎

The important point for us is that, as we saw, once we have a Division Algorithm, we have a Euclidean Algorithm and then we get results analogous to those we obtained for the ordinary integers and for polynomials in $F[x]$. First we need a definition.

Definition. Let m and n be nonzero Gaussian integers. A greatest common divisor of m and n is a Gaussian integer d such that

(a) d is a common divisor of m and n (i.e., there exist Gaussian integers a and b with $m = ad$ and $n = bd$); and

(b) If d_1 is any common divisor of m and n, then d_1 is a divisor of d.

We will write $d = gcd(m, n)$, although in this context it is an abuse of notation because d is not uniquely determined. Indeed if d is a gcd of m and n, then clearly so are $-d, id$, and $-id$. In any case the Euclidean Algorithm gives the existence of greatest common divisors exactly as before and yields the following versions of Euclid's Lemma.

Euclid's Lemma for $\mathbf{Z}[i]$ (Version 1). *Let m and n be nonzero Gaussian integers. Then m and n have a greatest common divisor d. Moreover there exist Gaussian integers a and b such that $d = am + bn$.*

Euclid's Lemma for $\mathbf{Z}[i]$ (Version 2). *Let m, a, and b be Gaussian integers. If m divides ab and $gcd(a, m) = 1$, then m divides b.*

Definitions.
(a) A Gaussian integer u is a **unit** in $\mathbf{Z}[i]$ if u has a multiplicative inverse in $\mathbf{Z}[i]$;
(b) Two Gaussian integers a and b are **associates** if $b = ua$ for some unit $u \in \mathbf{Z}[i]$;
(c) A Gaussian integer a is **irreducible** if a is not a unit and the only divisors of a are units and associates of a; and

(d) A Gaussian integer p is a **Gaussian prime** if whenever a and b are Gaussian integers with p dividing ab, then either p divides a or p divides b.

We saw before for an arbitrary ring R that the set $U(R)$ of units of R forms a group under the multiplication operation in R.

Exercises

14.3. Which Gaussian integers are units? Write the multiplication table for the unit group of $\mathbf{Z}[i]$.

14.4. Prove that the following definition is equivalent to the one given earlier.
A Gaussian integer γ is irreducible if γ is not a unit and γ cannot be written as the product $\gamma = \alpha\beta$ with $N(\alpha) < N(\gamma)$ and $N(\beta) < N(\gamma)$.

14.5. Verify that 2 and 5 are not irreducible Gaussian integers, but 3 and 7 are irreducible Gaussian integers.

14.6. Prove: A Gaussian integer m is an irreducible Gaussian integer if and only if m is a Gaussian prime. ∎

We now digress (apparently) to find elements of order 4 in the group U_p whenever $p \equiv 1 \pmod 4$. (Later we shall see another proof.)

Exercises

14.7. Prove: Let G be a finite abelian group with $|G|$ odd. Then the product of all the elements of G is the identity element e of G. (*Hint*: Since multiplication in G is commutative, the product may be taken in any order. Arrange the elements of G so that g is next to g^{-1}. Be careful: What if $g = g^{-1}$?)

14.8. Prove: If G is a finite abelian group with exactly one element f of order 2, then the product of all the elements of G is f.

14.9. Prove Wilson's Theorem: If p is a prime, then $(p-1)! \equiv -1 \pmod{p}$. (*Hint*: Apply 14.8 to the group U_p. Why does it apply?)

14.10. Let p be a prime with $p \equiv 1 \pmod 4$. By "splitting" $(p-1)!$ into two "equal parts," find an explicit formula for an integer a with $a^2 \equiv -1 \pmod p$.

14.11. Prove: If p is a prime, then U_p contains an element of order 4 (in fact two of them) if and only if $p \equiv 1 \pmod 4$. (*Hint*: Remember Lagrange's Theorem.)

14.12. Find the elements of order 4 in U_5, U_{13}, and U_{17}. ∎

We shall apply this to prove Fermat's Two Squares Theorem.

Fermat's Two Squares Theorem. *Let p be an odd prime integer. Then p is a sum of two squares if and only if $p \equiv 1 \pmod 4$.*

Proof. In one direction, the assertion is easy. Suppose that $p = a^2 + b^2$. Since p is odd, we may assume that $a = 2c$ and $b = 2d + 1$. Then
$$p = a^2 + b^2 = 4c^2 + 4d^2 + 4d + 1 = 4(c^2 + d^2 + d) + 1$$
and so $p \equiv 1 \pmod 4$.

The harder part is to show that *whenever* $p \equiv 1$ (mod 4), there exist integers a and b with $p = a^2 + b^2$. Here's the trick. Suppose that p is *not* a Gaussian Prime. Then we can write $p = (a + bi)(c + di)$ with $a + bi$ and $c + di$ not units; i.e., $N(a + bi) \neq 1$. But

$$p^2 = N(p) = N(a+bi)N(c+di) = (a^2+b^2)(c^2+d^2).$$

The only possible conclusion is that $p = a^2 + b^2 = c^2 + d^2$, and we are done. So we may assume that p is a Gaussian Prime.

Now the significance of the fact that $p \equiv 1$ (mod 4) is that by 14.10, there is an integer a with $a^2 \equiv -1$ (mod p). In other words there is a positive integer k such that

$$kp = a^2 + 1 = (a+i)(a-i).$$

Since p is a prime Gaussian integer, Euclid's Lemma says that p divides one of the two factors, say $a + i$. But then

$$a + i = p(b + ci) \text{ for some ordinary integers } b \text{ and } c.$$

In particular $pc = 1$, which is impossible, giving a final contradiction. ∎

This proves Fermat's celebrated theorem. Since it is such a lovely result, let's look at a slightly different way of completing the proof.

By analogy with $\mathbf{Z}/n\mathbf{Z}$ we can construct (for $n \in \mathbf{Z}$) some more rings, $\mathbf{Z}[i]/n\mathbf{Z}[i]$:

$$\mathbf{Z}[i]/n\mathbf{Z}[i] = \{a + bi : 0 \leq a, b \leq n - 1\}$$

with addition and multiplication done modulo n.

An interesting question is, When is the ring $\mathbf{Z}[i]/n\mathbf{Z}[i]$ a field?

Theorem. $\mathbf{Z}[i]/n\mathbf{Z}[i]$ *is a field if and only if n is a Gaussian Prime.*

Proof. Suppose first that n is not a Gaussian Prime, so that $n = \alpha\beta$ with α and β Gaussian integers such that $N(\alpha) < N(n)$ and $N(\beta) < N(n)$. Since $n \equiv 0$ (mod n), we have

$$\alpha\beta = n \equiv 0 \pmod{n}$$

and so

$$\alpha \cdot \beta = 0$$

in $\mathbf{Z}[i]/n\mathbf{Z}[i]$. If the Cancellation Law holds in $\mathbf{Z}[i]/n\mathbf{Z}[i]$, then either $\alpha = 0$ or $\beta = 0$ in $\mathbf{Z}[i]/n\mathbf{Z}[i]$. But if, for instance,

$$\alpha = na + nbi = n(a + bi)$$

for some integers a and b, not both 0, then

$$N(\alpha) = N(n)N(a + bi) \geq N(n),$$

contrary to assumption. Similarly for β. Hence $\alpha \neq 0 \neq \beta$ in $\mathbf{Z}[i]/n\mathbf{Z}[i]$ and so the Cancellation Law does not hold in $\mathbf{Z}[i]/n\mathbf{Z}[i]$ when n is not a Gaussian Prime.

On the other hand, suppose that n is a prime Gaussian integer and α is a Gaussian integer that is not a multiple of n. Then $gcd(\alpha, n) = 1$ and so by Euclid's Lemma, there exist Gaussian integers β and γ with

$$\beta\alpha + \gamma n = 1.$$

Thus

$$\beta \cdot \alpha = 1 - \gamma n \equiv 1 \pmod{n},$$

and so β is the multiplicative inverse of α in $\mathbf{Z}[i]/n\mathbf{Z}[i]$. Thus $\mathbf{Z}[i]/n\mathbf{Z}[i]$ is a field. ∎

Now recall that if F is any field of numbers and $p(x)$ is any polynomial of degree d with coefficients in F, then $p(x)$ has at most d roots in F. This gives us a somewhat different way of finishing the proof of Fermat's theorem.

Second proof of Fermat's theorem. We begin as in the first proof up to the point of concluding that we may assume that p is a prime Gaussian integer. Again we wish to derive a contradiction from this assumption.

Since p is a prime Gaussian integer, $F = \mathbf{Z}[i]/p\mathbf{Z}[i]$ is a field and so the polynomial $x^2 + 1$ has at most two roots in F.

On the other hand, since $p \equiv 1 \pmod{4}$, U_p contains an element a of order 4 by 14.10; i.e., there are ordinary integers a and $-a$ with $a^2 \equiv -1 \pmod{p}$.

Thinking of $\pm a$ as Gaussian integers, we see that $(\pm a)^2 \equiv -1$ in $\mathbf{Z}[i]/p\mathbf{Z}[i]$. But also obviously $(\pm i)^2 \equiv -1$ in $\mathbf{Z}[i]/p\mathbf{Z}[i]$, where i and $-i$ are the usual square roots of -1. On the other hand, no two of $a, -a, i$, and $-i$ are congruent modulo p in $\mathbf{Z}[i]$ (since $p \neq 2$) and so $x^2 + 1$ has at least four distinct roots in F, contrary to Descartes' Factor Theorem.

Thus p cannot be a prime Gaussian integer, proving Fermat's Theorem. ∎

Exercises

14.13. Write the addition and multiplication tables for the ring $\mathbf{Z}[i]/2\mathbf{Z}[i]$. Verify that there is a nonzero element a of this ring with $a^2 = 0$. Such an element is called **nilpotent**.

14.14. Write the addition and multiplication tables for the ring $F = \mathbf{Z}[i]/3\mathbf{Z}[i]$. Verify that F is a field by finding an inverse for each nonzero element of F.

14.15. I'm not sadistic enough to ask you to write the addition and multiplication tables for the ring $R = \mathbf{Z}[i]/5\mathbf{Z}[i]$. However, find an element e in R with $e \neq 0$ and $e \neq 1$, but $e^2 = e$. (*Hint*: Set $e = a + bi$. The equation $e^2 = e$ leads to a pair of simultaneous congruences modulo 5. Solve this system for a and b.) Such an element e is called **idempotent**. Notice that $e(1 - e) = 0$. Hence the Cancellation Law does not hold in R.

14.16. (Bonus) Prove: If p is a prime with $p \equiv 1 \pmod{4}$, then the ring $\mathbf{Z}[i]/p\mathbf{Z}[i]$ contains an idempotent element e with $e \neq 0$ and $e \neq 1$. Hence p is not a prime Gaussian integer. (This gives a third proof of Fermat's Theorem.)

Definition. We say that a natural number n is **square-free** if n has no divisor m^2 with $m \in \mathbf{N}$ and $m > 1$. (*Note*: If n is any natural number, then we may write $n = m^2 n_0$, where n_0 is a square-free natural number.)

14.17a. Using 14.1 and Fermat's Theorem, guess a rule for which square-free natural numbers n can be written as $n = a^2 + b^2$ with a and b natural numbers.

14.17b. Now guess a rule for which natural numbers n can be written as $n = a^2 + b^2$ with a and b non-negative integers. (*Note*: This includes the possibility that $n = a^2$ and $b = 0$.)

14.17c. The Pythagorean Theorem was close to Fermat's heart. Use 14.17a to give a rule for which natural numbers c can be the length of the hypotenuse of a right triangle whose sides have integer lengths a and b with $gcd(a, b) = 1$ (i.e., (a, b, c) is a primitive Pythagorean triple). Justify your rule.

14.18. (Bonus) Prove the validity of the rule you guessed in 14.17a. ∎

CHAPTER 14A

MORE RINGS AND MORE SQUARES

One of the few theorems of Fermat not proved by Euler was the Four Squares Theorem, for which the first published proof was given by Lagrange.

Four Squares Theorem. *Every natural number n can be written as the sum of four squares of nonnegative integers.*

The number $7 = 2^2 + 1^2 + 1^2 + 1^2$ shows that at least four squares are necessary. Adolf Hurwitz discovered a beautiful proof of the Four Squares Theorems using a subring of Hamilton's quaternions in a manner quite analogous to the way we used the Gaussian integers to prove the Two Squares Theorem.

Since the Hamilton quaternions **H** are a noncommutative ring, we shall need to be a bit more careful. First let's develop a little ring theory that will be useful. We shall only deal with rings that have a multiplicative identity element, 1.

Let $(R, +, \cdot)$ be a ring. Recall that in particular $(R, +)$ is an abelian group.

Definition. If the identity element 1 has finite order n in the group $(R, +)$, then we say that R has **characteristic** n. If 1 has infinite order, then we say that R has **characteristic** 0.

Definition. The **center** of R, $Z(R)$, is

$$Z(R) = \{z \in R : xz = zx \text{ for all } x \in R\}.$$

Exercises

14A.1. Prove: $Z(R)$ is a subring of R containing 1.

14A.2. Let Z be the cyclic subgroup of $(R, +)$ generated by 1. Prove:
 a. Z is a subring of $Z(R)$.
 b. $Z \cong \mathbf{Z}$ if and only if R has characteristic 0.
 c. $Z \cong \mathbf{Z}/n\mathbf{Z}$ if and only if R has characteristic n.

14A.3. Prove: If $Z(R)$ is a domain, then either R has characteristic 0 or R has characteristic p for some prime p. Give examples where each of these possibilities occurs.

14A.4. Prove: If R is a finite ring and $Z(R)$ is a domain, then $|R| = p^n$ for some prime p and $n \in \mathbf{N}$. (*Hint*: Prove that every nonzero element of $(R, +)$ has order p and apply Cauchy's Theorem.) ∎

In the theory of rings, a very important role is played by certain subrings called left (or right or 2-sided) ideals. They are analogous both to subspaces of a vector space and to subgroups of a group. In particular the so-called principal ideals correspond to one-dimensional subspaces of a vector space and to cyclic subgroups of a group. We shall only touch lightly on this large subject.

Definition. Let $(R, +, \cdot)$ be a ring.

(a) A subring $(I, +, \cdot)$ of R is called a **left ideal** of R if $ra \in I$ for all $r \in R$ and $a \in I$;
(b) A subring $(I, +, \cdot)$ of R is called an **ideal** or a **2-sided ideal** of R if both $ra \in I$ and $ar \in I$ for all $r \in R$ and $a \in I$;
(c) The **principal left ideal** containing a, Ra, is

$$Ra = \{ra : r \in R\};$$

(d) R is called a **principal (left) ideal ring** if every (left) ideal of R is principal.
Note that if R is a commutative ring, then we may and shall drop the adjective *left* in all the preceding terms.

The following construction will be of importance to us. It generalizes the construction of $\mathbf{Z}/n\mathbf{Z}$ and of $\mathbf{Z}[i]/p\mathbf{Z}[i]$. It only works for *ideals*, not for left ideals. The key fact is the following coset multiplication where I is an ideal in a ring R:

$$(a + I)(b + I) = ab + aI + Ib + I^2 \subseteq ab + I.$$

Note that it is necessary for I to be a 2-sided ideal in order to guarantee that both $aI \subseteq I$ and $Ib \subseteq I$.

Definition. Let $(R, +, \cdot)$ be a ring and let I be a (2-sided) ideal of R. The quotient ring $(R/I, +, \cdot)$ is the set of all cosets of I in R with the operations

$$(a + I) + (b + I) = (a + b) + I$$

and

$$(a + I)(b + I) = ab + I.$$

Note that the set multiplication given earlier,

$$(a + I)(b + I) \subseteq ab + I,$$

guarantees that the product of the cosets $a + I$ and $b + I$ is uniquely determined independent of the choice of coset representatives a and b. In other words, the product could have been defined by this rule:

If J and K are two cosets of I, then the product $J \cdot K$ is the unique coset of I containing the set JK.

This definition avoids all reference to coset representatives. Once you are convinced that addition and multiplication in R/I is well defined, it is clear that these operations inherit any associative, commutative, and distributive laws that hold in R itself. However, it is not true that R/I will in general inherit the Cancellation Law for Multiplication from R. For example, $\mathbf{Z}$ is a domain, but $\mathbf{Z}/n\mathbf{Z}$ is not a domain whenever n is a composite number.

Exercises

14A.5. Prove Ra is a left ideal of R.

14A.6. Let $\mathbf{Z}[x]$ be the ring of polynomials with integer coefficients. Let I be the ideal generated by the polynomials 2 and x, i.e., the smallest ideal of $\mathbf{Z}[x]$ containing both 2 and x. Describe the polynomials in I. Prove that I is not a principal ideal. ∎

The most effective tool for proving that ideals are principal is the Division Algorithm.

Theorem. *Let $R = \mathbf{Z}$ or $\mathbf{Z}[i]$ or $F[x]$, for some field F. Then R is a principal ideal ring.*

Proof. Let I be an ideal in R. If $I = \{0\}$, then clearly $I = R0$ is principal. If not, choose d in I of minimal size subject to $d \neq 0$. (Here *size* means absolute value or norm or degree, according to the case.) Now if $a \in I$, we apply the Division Algorithm to obtain

$$a = qd + r,$$

where either $r = 0$ or r is of smaller size than d. But $r = a - qd$ is in I. So r cannot be of smaller size than d. Hence $r = 0$ and $a = qd \in Rd$. Thus $I = Rd$ is a principal ideal. ∎

By an obvious generalization we obtain the following theorem.

Theorem. *Let R be a ring having a Left Division Algorithm. Then R is a principal left ideal ring.*

A key ingredient in Hurwitz' proof is the construction of a subring of $\mathbf{H}$ having a Left Division Algorithm. But before taking that step, let's do a few more general preliminaries.

Recall that a ring R with 1 is a **skew field** or **division ring** if every nonzero element of R has a multiplicative inverse in R; i.e., if $(R - \{0\}, \cdot)$ is a group. When R is a skew field, we shall write $R^\times$ to denote the group $(R - \{0\}, \cdot)$.

Lemma. *Let R be a ring with 1. Then R is a skew field if and only if $Ra = R$ for every $a \in R^\times$.*

Proof. If R is a skew field and a is a nonzero element of R, then a has a multiplicative inverse b. Then for every $r \in R$,

$$r = r \cdot 1 = r(ba) = (rb)a \in Ra,$$

and so $Ra = R$, as claimed.

Next suppose that $Ra = R$ for all nonzero $a \in R$. Let $a \in R^\times$. Since $Ra = R$, there exists $b \in R$ with $ba = 1$. We must show also that $ab = 1$. Since multiplication in R is not necessarily commutative, this is not obvious. However $b \neq 0$ and so $Rb = R$. Hence there exists $c \in R$ with $cb = 1$. Now we again use the Associative Law for Multiplication in R:

$$c = c \cdot 1 = c(ba) = (cb)a = 1 \cdot a = a.$$

Hence $ab = 1$, as desired. ∎

In 1905, J. H. M. Wedderburn proved the following celebrated theorem.

Wedderburn's Theorem. *Let R be a finite skew field. Then R is a field; i.e., multiplication in R satisfies the Commutative Law.*

The proof of this theorem would take us a bit far afield. Instead we shall do just what we need for Hurwitz' proof. It will give you something of the flavor of the general proof.

Definition. Let R be a ring and let $a \in R$. The **centralizer** of a in R or the **commuting ring** of a in R is

$$C_R(a) = \{x \in R : ax = xa\}.$$

Thus $C_R(a) = R$ if and only if $a \in Z(R)$. Also $Z(R) \leq C_R(a)$ for all $a \in R$.

Exercises

14A.7a. Prove: $C_R(a)$ is a subring of R for all $a \in R$.

14A.7b. Prove: If R is a skew field, then $C_R(a)$ is also a skew field.

14A.8. Prove: If $Z(R) = C_R(a)$ for some $a \in R$, then $Z(R) = R$.

14A.9. Prove: Let $a, m, n \in \mathbf{N}$ with $a > 1$. Then $a^m - 1$ divides $a^n - 1$ if and only if m divides n. (Compare with 11.19)

14A.10. Let R be a finite skew field.

14A.10a. Prove: $|R| = p^n$ for some prime p and $n \in \mathbf{N}$.

14A.10b. Prove: If S is a subskew field of R, then $|S| = p^m$ for some $m \in \mathbf{N}$ with m a divisor of n.

14A.10c. Prove: If n is a prime, then R is a field. ∎

The exercises have established that any finite skew field of cardinality p, p^2, or p^3 for p a prime must be a field. The first unsettled case is that of skew fields of cardinality p^4. This is precisely the case we shall need for Hurwitz' proof. It follows easily from Cauchy's Theorem.

Theorem. *Let R be a finite skew field with $|R| = p^4$ for some prime p. Then R is a field.*

Proof. Suppose that R is not a field and let a be any element of $R - Z(R)$. By the preceding exercises, $C_R(a)$ is a subskew field of R with $Z(R) < C_R(a) < R$ and with $|C_R(a)| = p^m$

where m is a divisor of 4. The only possibility is that $|Z(R)| = p$ and $|C_R(a)| = p^2$ for all $a \in R - Z(R)$.

If $p = 2$, set $r = 5$. If p is odd, then $gcd(p^2 - 1, p^2 + 1) = 2$ and $p^2 + 1 = 2m$ for some odd $m > 1$. In this case, let r be a prime divisor of m. By Cauchy's Theorem, since $|R^\times| = p^4 - 1$ is divisible by r, $R^\times$ contains an element a of order r. If $a \in Z(R)^\times$, then by Lagrange's Theorem, r divides $|Z(R)^\times| = p - 1$, which is not the case. Thus $a \notin Z(R)$ and so $|C_R(a)| = p^2$. But then $a \in C_R(a)^\times$ and so again by Lagrange's Theorem, r divides $p^2 - 1$, which is not the case. This contradiction completes the proof. ∎

The proof of the complete theorem of Wedderburn follows (almost completely) by the identical argument, in conjunction with the following theorem of Zsigmondy.

Zsigmondy's Theorem. *Let p be a prime and n a natural number with $n > 2$. There is always a prime r that is a divisor of $p^n - 1$ but is not a divisor of $p^m - 1$ for any divisor m of n with $m < n$, with one exception: $p^n = 2^6$.*

Exercises

14A.11. Using Zsigmondy's Theorem, prove that every finite skew field D is a field, unless perhaps $|D| = 64$.

14A.12. (Bonus) Prove: Every skew field D with $|D| = 64$ is a field. (*Hint*: If $a \in D^\times$ has order 7, then so does xax^{-1} for all $x \in D^\times$. Deduce that the number of elements of D of order 7 is a multiple of 9. On the other hand, by a sharp version of Cauchy's Theorem, the number of subgroups of $D^\times$ of order 7 is $7k + 1$ for some $k \geq 0$. Derive a contradiction.) ∎

In the Appendix to Chapter 19, we will give a proof of Zsigmondy's theorem for odd primes p.

Now we turn to Hurwitz' ring J of integral quaternions. The obvious ring of integral quaternions is

$$J_0 = \{a + bi + cj + dk : a, b, c, d \in \mathbf{Z}\}.$$

However J_0 is just a bit too small for the Division Algorithm to work.

Imagine that we are trying to give a geometric proof of a Left Division Algorithm for J_0 analogous to the proof for $\mathbf{Z}[i]$. Thus given $m, n \in J_0$, we are trying to find q, $r \in J_0$ with

$$n = qm + r$$

and

$$|r| < |m|,$$

where $|\ |$ is the usual Euclidean length function in $\mathbf{R}^4$. As before we observe that $\mathbf{Z}m$ is the set of integer multiples of m lying along the line $\mathbf{R}m$ in $\mathbf{R}^4$. And again, left multiplication by i, j, or k rotates this line 90°, each time in a different direction, so we end up with a set of four mutually perpendicular axes in $\mathbf{R}^4$ and the J_0-multiples of m determine a set of lattice points at the vertices of four-dimensional hypercubes in $\mathbf{R}^4$ with all sides of length $|m|$. Again we look at the hypercube containing n and choose

qm to be a vertex closest to n. It remains to argue that

$$|r| = |n - qm| < |m|,$$

i.e., that the distance of n from a nearest vertex qm is less than the length of a side of the hypercube. Again the "worst-case scenario" is clearly when n is at the center of the hypercube. Then it is equidistant from every vertex and the distance is

$$\sqrt{\left(\frac{m}{2}\right)^2 + \left(\frac{m}{2}\right)^2 + \left(\frac{m}{2}\right)^2 + \left(\frac{m}{2}\right)^2} = m.$$

Damn! So we fail at exactly one point in each hypercube, namely the point

$$(q + \zeta)m,$$

where $\zeta = \frac{1}{2}(1 + i + j + k)$ and qm is one of the vertices of the hypercube. A solution leaps to mind: Enlarge J_0 slightly so that it contains all of these points.

Definition. The ring J of Hurwitz' integral quaternions is the ring

$$J = J_0 \cup (\zeta + J_0) = \{a\zeta + bi + cj + dk : a, b, c, d \in \mathbf{Z}\}.$$

Exercises

14A.11. Verify that

$$J_0 \cup (\zeta + J_0) = \{a\zeta + bi + cj + dk : a, b, c, d \in \mathbf{Z}\}.$$

Recall that if $\alpha = a + bi + cj + dk \in \mathbf{H}$, then $\alpha^* = a - bi - cj - dk$ and

$$N(\alpha) = \alpha\alpha^* = a^2 + b^2 + c^2 + d^2.$$

14A.12a. Verify that J is a subring of $\mathbf{H}$

14A.12b. Verify that $\alpha^* \in J$ whenever $\alpha \in J$.

14A.12c. Verify that $N(\alpha)$ is a nonnegative integer for all $\alpha \in J$.

14A.12d. Verify that $\alpha^* = \alpha^{-1}$ if $N(\alpha) = 1$.

14A.13. Prove: Let $\alpha \in J$. Then the left ideal $J\alpha = J$ if and only if $N(\alpha) = 1$. (*Hint: The norm map is multiplicative.*) ∎

Now we have enlarged J_0 little enough so that the norm map N defines a size function on J taking values in the nonnegative integers. On the other hand we have added enough elements to avoid the embarrassment with the centers of hypercubes and so clearly we have a Left Division Algorithm in J and the following theorem holds.

Theorem. *Every left ideal of J is a principal left ideal.*

We now have all the tools we need to prove the Four Squares Theorem. First we observe the following:

Lemma. *Let $n \in \mathbf{N}$.*

(a) *If $n = ab$ with $a, b \in \mathbf{N}$ and both a and b are sums of four squares, then also n is a sum of four squares; and*

(b) *n is a sum of four squares if and only if* $n = N(\alpha)$ *for some* $\alpha \in J$.

Proof. Clearly if a and b are sums of four squares, then there exist $\alpha, \beta \in J_0$ with $a = N(\alpha)$ and $b = N(\beta)$. But then

$$n = ab = N(\alpha)N(\beta) = N(\alpha\beta),$$

and $\alpha\beta \in J_0$, so n is also a sum of four squares, proving (a).

Now if $n = N(\alpha)$ with $\alpha \in J$, then $2\alpha \in J_0$ and

$$N(2\alpha) = N(2)N(\alpha) = 4n.$$

So $4n$ is a sum of four squares. We now use a trick of Euler. Suppose

$$2m = a^2 + b^2 + c^2 + d^2,$$

with a, b, c, d nonnegative integers. Then either 0, 2, or 4 of these integers must be odd and so we may relabel them so that

$$A = \frac{a+b}{2}, B = \frac{a-b}{2}, C = \frac{c+d}{2}, D = \frac{c-d}{2}$$

are nonnegative integers. But then we easily check that

$$m = A^2 + B^2 + C^2 + D^2.$$

Using this trick twice, we see first that $2n$ is a sum of four squares and then that n is a sum of four squares, as claimed. ∎

Now we can prove

The Four Squares Theorem. *Let* $n \in \mathbf{N}$. *Then*

$$n = a^2 + b^2 + c^2 + d^2$$

for some nonnegative integers a, b, c, and d.

Proof. Clearly the theorem is true for $n = 1$ and $n = 2$. By the previous lemma, if the theorem is also true for all odd primes, then it is true for all $n \in \mathbf{N}$, since n may be factored as a product of primes.

Thus it will suffice to show that if p is an odd prime, then $p = N(\alpha)$ for some $\alpha \in J$. Suppose that there is a proper left ideal I in J with Jp properly contained in I. Then I must be a principal left ideal and so $I = J\alpha$ for some $\alpha \in J$. Since $p \in I$, there exists $c \in J$ with $p = c\alpha$. If $N(c) = 1$, then $c^* = c^{-1} \in J$ and $\alpha = c^*p \in Jp$, contrary to the assumption that $Jp \neq I$. Hence $N(c) > 1$. If $N(\alpha) = 1$, then $J\alpha = J$, again contrary to assumption. Hence also $N(\alpha) > 1$. But

$$p^2 = N(p) = N(c\alpha) = N(c)N(\alpha)$$

and so $N(c) = N(\alpha) = p$, and we are done!

Thus we may assume that there is no left ideal I in J that is strictly bigger than Jp, except for J itself.

Now a very important point is that p commutes with every element of J and so $Jp = pJ$ is a 2-sided ideal of J. This means that, in a manner completely analogous to the construction of the rings $\mathbf{Z}/n\mathbf{Z}$ and $\mathbf{Z}[i]/p\mathbf{Z}[i]$, we may construct the ring

$$J/pJ = \{a\zeta + bi + cj + dk : a, b, c, d \in \mathbf{Z}/p\mathbf{Z}\}.$$

(Notice that since p is odd,

$$\frac{1}{2} = \frac{p-1}{2} \in \mathbf{Z}/p\mathbf{Z}$$

and so we may also write

$$J/pJ = \{a + bi + cj + dk : a, b, c, d \in \mathbf{Z}/p\mathbf{Z}\}.)$$

In any case, we observe that J/pJ is a finite ring with $|J/pJ| = p^4$. Furthermore $ij \neq ji \in J/pJ$, again since $p \neq 2$. So J/pJ is not a field. Hence by our special case of Wedderburn's Theorem, J/pJ is not a skew field and so there exists a left ideal I/pJ of J/pJ that is neither the whole ring nor the zero ring. But then the full pre-image I in J is a left ideal of J that is strictly bigger than pJ but is not the whole ring J, contrary to assumption.

This completes Hurwitz' proof of the Four Squares Theorem. ∎

CHAPTER 15

FERMAT'S LAST THEOREM (FOR POLYNOMIALS)

Cujus rei demonstrationem mirabilem sane detexi.
Hanc marginis exiguitas non caparet.

[I have discovered a truly marvelous proof of this, which this margin is too narrow to contain.]

—Pierre Fermat

(In the margin of his copy of Bachet's annotated translation of the "Arithmetic" of Diophantus)

Fermat's most famous "theorem" is the one he probably never proved, namely:

Fermat's Last Theorem. *For any $n \geq 3$, there is no triple (a, b, c) of natural numbers satisfying*

$$a^n + b^n = c^n.$$

This statement has a long and illustrious history. The most dramatic moment of modern mathematics was the proof of this theorem by Andrew Wiles, completed in 1994. The proof is incredibly long and deep. Indeed Wiles' contribution does not directly address Fermat's Last Theorem at all. Rather it establishes a major case of a conjecture of Taniyama about elliptic curves. By a theorem of Frey, Serre, and Ribet (established shortly *before* Wiles' work), this theorem of Wiles implies the truth of Fermat's Last Theorem.

Remarkably there is a version of Fermat's Last Theorem for polynomials that is not too hard to prove. This is further evidence of the existence of many analogies between numbers and polynomials. It also supports the assertion that results are usually much easier to prove for polynomials than they are for integers. We shall give the proof in this section.

Recall that for polynomials, as for integers, it makes sense to talk about $gcd(f, g)$ and so we may call two polynomials f and g relatively prime if $gcd(f, g) = 1$.

Fermat's Last Theorem for Polynomials. *For any $n \geq 3$, there is no triple (f, g, h) of polynomials with complex coefficients satisfying*

$$f^n + g^n = h^n$$

with $gcd(f, g) = 1$ and at least one of the polynomials nonconstant.

Notice that since every complex number has a complex nth root for all n, if f and g are constant polynomials, then we can always find a constant polynomial h (in fact n of them) making the Fermat equation hold. So we certainly have to assume that at least one of the polynomials is nonconstant. Furthermore the *gcd* condition is just there to prevent us from taking a trivial constant polynomial solution and multiplying it through on both sides by some nonconstant polynomial f to get an (almost equally trivial) solution; e.g., if α is a cube root of 9, then the trivial equation

$$1^3 + 2^3 = \alpha^3$$

yields the equally trivial solution

$$x^3 + (2x)^3 = (\alpha x)^3.$$

Exercise

15.1. Use the formula for Pythagorean triples to show that there are nontrivial polynomial solutions of

$$f^2 + g^2 = h^2.$$ ∎

The proof of this theorem for polynomials is interesting because to those who don't believe that Fermat actually had a correct proof of his "Last Theorem," it suggests a possible incorrect but plausible proof that might have fooled Fermat and indicates one reason this problem is so subtle.

The proof proceeds by Fermat's favorite method, the so-called Method of Infinite Descent. If the theorem is false, then there is a triple of not-all-constant polynomials

f, g, and h with $\gcd(f, g) = 1$ satisfying

$$f^n + g^n = h^n.$$

The "descent" involves producing a second triple of again non-all-constant polynomials F, G, H with $\gcd(F, G) = 1$ satisfying

$$F^n + G^n = H^n$$

and *smaller* than the original triple in some sense of the word *smaller*. We can measure smallness in a variety of ways. Let's say that (F, G, H) is smaller than (f, g, h) if the maximum of the degrees of G and H is less than the maximum of the degrees of g and h.

The descent must proceed in such a way that it can be repeated arbitrarily often (hence "infinite descent"). On the other hand, since the original maximum degree is a finite positive integer and the maximum is necessarily a positive integer at every step, the descent must stop after *finitely* many steps, and this contradiction proves that it could never have started!

OK. We begin by rewriting the equation as

$$f^n = h^n - g^n$$

and then factoring the right side of the equation. Namely we let ζ be a primitive nth root of 1. Then

$$f^n = h^n - g^n = \prod_{j=0}^{n-1}(h - \zeta^j g).$$

Exercise

15.2. Prove this. (*Hint*: Write $h/g = x$ and factor $x^n - 1$.) ■

Now the key point of the proof is the following assertion:

For each j, $h - \zeta^j g$ is an nth power.

That is, there exist polynomials k_j for $0 \leq j \leq n - 1$ such that

$$h - \zeta^j g = k_j^n.$$

Once we prove this statement it will be easy to "descend" to a smaller solution:

$$k_2^n + (\alpha k_0)^n = (\beta k_1)^n$$

for suitably chosen complex numbers α and β, and then we are done.

So why should it be true that $h - \zeta^j g$ is an nth power? The reason is that since $\gcd(g, h) = 1$, it follows that the polynomials $h - \zeta^j g$ are all relatively prime, but they multiply up to $f^n = ff \cdots f$, a product of n *equal* polynomials. So every prime factor of f must occur n times on the left and hence n times in *exactly one* factor on the right.

For example in the integers, if we factor 6^2 into relatively prime factors, the only way we can do it is if each of the factors is itself a perfect square:

$$6^2 = 4 \cdot 9 = 2^2 \cdot 3^2.$$

This is a consequence of the Unique Factorization Theorem for the integers. A similar statement holds whenever we have a Unique Factorization Theorem. But we do have a Unique Factorization Theorem for Polynomials in $\mathbf{C}[x]$. And so in consequence we obtain the following result.

Theorem. *Let $a(x)$, $b(x)$, and $c(x)$, be nonzero polynomials in $\mathbf{C}[x]$ with $\gcd(a(x), b(x)) = 1$ and with $c(x)^n = a(x)b(x)$. Then there exist polynomials $f(x)$ and $g(x)$ in $\mathbf{C}[x]$ with $a(x) = f(x)^n$ and $b(x) = g(x)^n$.*

Proof. By the Unique Factorization Theorem for Polynomials we may write

$$a(x) = k_a p_1(x)^{n_1} \cdots p_r(x)^{n_r}$$

and

$$b(x) = k_b q_1(x)^{m_1} \cdots q_s(x)^{m_s}$$

where the $p_i(x)$'s and the $q_i(x)$'s are distinct monic nonconstant irreducible polynomials in $\mathbf{C}[x]$, and both k_a and k_b are nonzero complex numbers. Since $\gcd(a, b) = 1$, no $p_i(x)$ can equal $q_j(x)$ for any i and j. So by the Unique Factorization Theorem again

$$c(x)^n = a(x)b(x) = (k_a k_b) p_1(x)^{n_1} \cdots p_r(x)^{n_r} q_1(x)^{m_1} \cdots q_s(x)^{m_s},$$

with the right-hand side being the unique (up to the ordering of the factors) factorization of $c(x)^n$ into irreducible polynomial factors.

On the other hand, let's write

$$c(x) = k_c f_1(x)^{h_1} \cdots f_t(x)^{h_t},$$

where again the f_i's are distinct monic nonconstant irreducible polynomials in $\mathbf{C}[x]$ and k_c is a nonzero complex number. Then we have a second factorization for $c(x)^n$, namely

$$c(x)^n = k_c^n f_1(x)^{h_1 n} \cdots f_t(x)^{h_t n}.$$

By uniqueness, the sets

$$\{p_1(x)^{n_1}, \ldots, p_r(x)^{n_r}, q_1(x)^{m_1}, \ldots, q_s(x)^{m_s}\}$$

and

$$\{f_1(x)^{h_1 n}, f_2(x)^{h_2 n}, \ldots, f_t(x)^{h_t n}\}$$

are identical sets and so, after suitable reordering of the second set, we may assume for example that

$$p_1(x)^{n_1} = f_1(x)^{h_1 n}, \ldots, q_s(x)^{m_s} = f_t(x)^{h_t n}.$$

In particular, for all i and j,

$$n_i = h_i n \text{ and } m_j = h_{r+j} n.$$

Thus

$$a(x) = c_a (p_1(x)^{h_1} \cdots p_r(x)^{h_r})^n$$

and

$$b(x) = c_b (q_1(x)^{h_{r+1}} \cdots q_s(x)^{h_{r+s}})^n.$$

Since we may take nth roots in $\mathbf{C}$, we are done by taking
$$f(x) = c_a^{1/n} p_1(x)^{h_1} \cdots p_r(x)^{h_r}$$
and
$$g(x) = c_b^{1/n} q_1(x)^{h_{r+1}} \cdots q_s(x)^{h_{r+s}}.$$
(Of course there are n different equally good choices for $c_a^{1/n}$ and likewise for $c_b^{1/n}$.) ∎

In order to apply this result to the proof of Fermat's Last Theorem for Polynomials, we need one more easy result.

Theorem. *Let $h(x)$ and $g(x)$ be polyonomials in $\mathbf{C}[x]$ with $\gcd(h, g) = 1$. Let α and β be distinct complex numbers. Then $\gcd(h + \alpha g, h + \beta g) = 1$.*

Proof. Suppose that p is a polynomial that divides both $h + \alpha g$ and $h + \beta g$. Then p divides their difference, which is $(\alpha - \beta)g$. Since $\alpha \neq \beta$, p divides g (and of course p divides αg). Then also p divides $(h + \alpha g) - \alpha g = h$. Since p divides both h and g, p is a nonzero constant polynomial and so $\gcd(h + \alpha g, h + \beta g) = 1$, as claimed. ∎

Exercises

15.3. Prove: If n is an integer with $n \geq 2$, then $\gcd(n - 1, n + 1) = 1$ or 2. What property of n determines which is the case?

15.4. Similarly, what can you say about $\gcd(n - 1, n + 2)$? ∎

Now we complete the proof of Fermat's Last Theorem for Polynomials. We have the equation
$$f^n = h^n - g^n = \prod_{j=0}^{n-1}(h - \zeta^j g).$$

Since h and g are relatively prime polynomials, no two of the polynomials on the right-hand side have a nonconstant common factor and so by an obvious extension of our theorem we get that there exist polynomials $k_j(x) \in \mathbf{C}[x]$, $0 \leq j \leq n - 1$, with
$$h - \zeta^j g = k_j^n$$
for all j.

Now we use the following simple formula:
$$(h - \zeta^2 g) + \zeta(h - g) = (1 + \zeta)(h - \zeta g).$$
(Notice that this is the one place where we use the fact that $n \geq 3$.) Since $h - \zeta^j g = k_j^n$, this translates into
$$k_2^n + \zeta k_0^n = (1 + \zeta)k_1^n.$$

Now every complex number has a complex nth root and so we can find numbers α and β with $\alpha^n = \zeta$ and $\beta^n = 1 + \zeta$. So
$$k_2^n + (\alpha k_0)^n = (\beta k_1)^n$$

and we have our "descent equation." Now we just have to check that the new triple $(k_2, \alpha k_0, \beta k_1)$ satisfies all the hypotheses of the theorem and is "smaller" than the original triple (f, g, h).

First of all, since the $h - \zeta^j g$'s are relatively prime, so are the k_j's. Also the degree of $h - \zeta^j g$ is at most the maximum of the degrees of g and h. Then the degree of k_j is $1/n$ the degree of $h - \zeta^j g$, hence smaller than this maximum, for all j. Thus the triple $(k_2, \alpha k_0, \beta k_1)$ really is smaller than (f, g, h).

But is it *too* small? Could they all be constants? Suppose that $h - g$ is a constant polynomial. Now if h and g are both constants, then so is f, contrary to assumption. So either h or g is not constant. Since $h - g$ is constant, both h and g must have the same degree $d > 0$ and we can write

$$h = ax^d + \cdots$$

and

$$g = bx^d + \cdots$$

Since $h - g$ is constant, $a = b \neq 0$. But then

$$h - \zeta g = a(1 - \zeta)x^d + \cdots$$

and so $h - \zeta g$ is not a constant polynomial. Thus $(k_2, \alpha k_0, \beta k_1)$ satisfies all of the hypotheses of the theorem and is smaller than (f, g, h).

So we have accomplished our descent and the theorem is proved!

Of course this does not prove the *real* Fermat's Last Theorem. Even the proof that

$$x^3 + y^3 = z^3$$

has no positive integer solutions is harder than the polynomial case, as you may see by reading the next chapter. Nevertheless the proof of this case and the early results on Fermat's Last Theorem up through the work of Kummer in the 1840s proceeded in this spirit. One would factor $z^n - y^n$ in the domain of cyclotomic integers: $\mathbf{Z}[\zeta]$, where $\zeta = e^{2\pi i/n}$. The main problem is this:

The Unique Factorization Theorem may be *false* in this domain.

Exercises

15.5a. Prove that the norm map on

$$\mathbf{Z}[\sqrt{-5}] = \{a + b\sqrt{-5} : a, b \in \mathbf{Z}\}$$

takes nonnegative integer values and satisfies

$$N(\alpha\beta) \geq N(\alpha)N(\beta) \text{ for all } \alpha, \beta \in \mathbf{Z}[\sqrt{-5}].$$

15.5b. List the four smallest nonzero norms of numbers in $\mathbf{Z}[\sqrt{-5}]$.

15.6. Prove that 1 and -1 are the only units in $\mathbf{Z}[\sqrt{-5}]$.

15.7. Prove that if $d \in \mathbf{Z}[\sqrt{-5}]$ and $1 < N(d) < 16$, then d is a prime in $\mathbf{Z}[\sqrt{-5}]$.

15.8. Using small primes, give an example of failure of unique factorization in $\mathbf{Z}[\sqrt{-5}]$. ∎

The problem seems to be that the norm function on $\mathbf{Z}[\sqrt{-5}]$ doesn't take enough small values and so we end up with too many primes.

Definition. A domain in which the Unique Factorization Theorem holds is called a **unique factorization domain** or **UFD**.

Several excellent mathematicians, including Lamé and Cauchy (and possibly Fermat) stumbled into false proofs of Fermat's Last Theorem by tacitly assuming that Unique Factorization would always hold. Kummer realized that a Unique Factorization Theorem does not always hold in this setting, just as it does not hold in $\mathbf{Z}[\sqrt{-5}]$. He investigated exactly when it does hold and was able to prove Fermat's Last Theorem for those values of n. However since Unique Factorization often fails, Kummer's methods were not destined to yield a complete proof of Fermat's Last Theorem. This would have to wait for another 150 years and for a completely different approach to the problem. Nevertheless Kummer's investigation of unique factorization in cyclotomic domains was one of the most fruitful and influential studies in the history of number theory. By analyzing and clarifying the work of Gauss, Dirichlet, and Kummer, Richard Dedekind laid the foundations for the general theory of ideals in commutative rings.

Exercises

Fermat's Last Theorem for $n = 4$

15.9. Prove: In order to prove Fermat's Last Theorem, it suffices to handle the cases $n = 4$ and n is an odd prime. ∎

It is certain that Fermat proved the following theorem:

Theorem (Fermat). *There is no solution in positive integers for*

$$x^4 + y^4 = z^2.$$

He used his method of infinite descent. We shall follow in his footsteps.

15.10. Prove: If x is a square, then $x \equiv 0$ or $1 \pmod 4$. ∎

Let's revisit primitive Pythagorean triples. Recall that the triple of positive integers (x, y, z) is called a primitive Pythagorean triple if

$$x^2 + y^2 = z^2$$

and $gcd(x, y) = 1$.

15.11. Prove: If (x, y, z) is a primitive Pythagorean triple, then exactly one of x and y is even. ∎

Now as in the proof of Fermat's Last Theorem for Polynomials, let's factor the left-hand side in $\mathbf{Z}[i]$:

$$(x + iy)(x - iy) = z^2.$$

15.12. Prove: $gcd(x + iy, x - iy) = 1$. (*Hint*: It is easy to see that, if not, then $gcd(x + iy, x - iy) = 1 + i$. But in this case argue that $x^2 + y^2$ is even, contrary to 15.11.)

15.13a. Prove: There exists a Gaussian integer $r + si$ such that either $x + iy = (r + si)^2$ or $x + iy = i(r + si)^2$.

15.13b. Using 15.13a, get the usual formulas for primitive Pythagorean triples.

15.14. Use the results on primitive Pythagorean triples to show that if (x, y, z) is a primitive solution of
$$x^4 + y^4 = z^2,$$
then there are positive integers u and v with u odd and v even and $gcd(u, v) = 1$ such that
$$x^2 = u^2 - v^2, \ y^2 = 2uv, \ z = u^2 + v^2.$$
(*Be careful*: u and v do not play symmetrical roles in these formulas. So you have to rule out the possibility that u is even and v is odd.)

15.15. Prove: There are positive integers r and s with
$$u = r^2, \ \frac{v}{2} = s^2, \text{ and } x^2 + 4s^4 = r^4.$$

15.16. Apply the theorem on primitive Pythagorean triples one more time to the last equation and use the same trick as in 15.14 to get at last an equation,
$$a^4 + b^4 = r^2,$$
for some positive integers a and b. Conclude that we are done by Fermat's method of infinite descent. Explain carefully.

15.17. Explain why you have proved Fermat's Last Theorem for $n = 4$. ∎

CHAPTER 15A

STILL MORE FERMAT'S LAST THEOREM (OPTIONAL)

On August 4, 1753, in a letter to Goldbach, Euler announced a proof of Fermat's Last Theorem for the case $n = 3$. He finally published a proof in 1770 in his algebra textbook. The proof was essentially correct, though not quite correct. In this section we sketch a correct proof.

We let $\omega = e^{2\pi i/3}$, a primitive cube root of 1. Thus the three cube roots of 1 are 1, ω, and ω^2, and they are linked by the equation

$$1 + \omega + \omega^2 = 0.$$

We define

$$\mathbf{Z}[\omega] = \{a + b\omega : a, b \in \mathbf{Z}\}.$$

Exercises

15A.1. Prove: $\mathbf{Z}[\omega]$ is a subdomain of $\mathbf{C}$; i.e., it is closed under addition and multiplication.

15A.2. Prove: The norm $N(a + b\omega) = a^2 - ab + b^2$. Find all of the units in $\mathbf{Z}[\omega]$. List all the elements of $\mathbf{Z}[\omega]$ of norm at most 3.

Definition. $\lambda = 1 - \omega$.

15A.3. Prove: λ^2 is an associate of 3. (In particular, 3 is not a prime.) ∎

As usual, in order to prove a Unique Factorization Theorem in $\mathbf{Z}[\omega]$, it suffices to prove a Division Algorithm. We may proceed geometrically as with the Gaussian integers. This time the fundamental lattice is composed of rhombuses with angle 120° and 60°, instead of squares.

165

Exercises

15A.4. Give an argument that in a rhombus of side a, every point is at a distance less than a from some vertex. ∎

This exercise is what is needed to establish the Division Algorithm and hence the Unique Factorization Theorem for $\mathbf{Z}[\omega]$.

15A.5. Explain why the Division Algorithm for $\mathbf{Z}[\omega]$ follows from 15A.4. ∎

The number λ behaves a lot like 3 in our ring. In particular we have the following lemma.

Lemma. *Every number in $\mathbf{Z}[\omega]$ is congruent to 0, 1, or -1 modulo λ. Thus λ divides $(\alpha - 1)\alpha(\alpha + 1)$ for all $\alpha \in \mathbf{Z}[\omega]$.*

Proof. If $\alpha = a + b\omega$, then $\alpha = (a + b) - b\lambda$ and so α is congruent to $a + b$ modulo λ. Since 3 divides λ, the result follows. ∎

From this we get the following little fact.

Lemma. *If α is not divisible by λ, then*
$$\alpha^3 \pm 1 \text{ is divisible by } \lambda^4.$$

Sketch of a proof. Replacing α by $-\alpha$ if necessary, we may write
$$\alpha = 1 + \beta\lambda.$$
With a little computation we get
$$\alpha^3 - 1 = \lambda^3 \beta(\beta + 1)(\beta - \omega^2).$$
Since ω^2 is congruent to 1 modulo λ, the result follows, using the previous lemma. ∎

Corollary. *If x, y, and z are in $\mathbf{Z}[\omega]$ with*
$$x^3 + y^3 + z^3 = 0,$$
then one of x, y, or z is divisible by λ.

Proof. If not, then by the previous lemma, 0 is congruent to $\pm 1 \pm 1 \pm 1$, modulo λ^4. So λ^4 divides $\pm 1 \pm 1 \pm 1$. But $N(\lambda^2) = N(3) = 9$. So $N(\lambda^4) = 81$ and λ^4 can't divide any number of norm less than 81. ∎

Now we are ready to formulate the version of Fermat's Last Theorem for $n = 3$, which we will prove by Fermat's Method of Infinite Descent.

FLT for $n = 3$. *There is no solution in $\mathbf{Z}[\omega]$ for*
$$x^3 + y^3 + \epsilon \lambda^{3m} z^3 = 0$$
where $\gcd(x, y) = 1$, ϵ is a unit, $m \geq 1$, and λ does not divide x, y, or z.

Exercise

15A.6. Explain why this will give a proof of the usual FLT for $n = 3$. ∎

First we argue that there is no solution when $m = 1$.

Lemma. *The result is true for $m = 1$.*

Proof. As neither x nor y is divisible by λ, the lemma tells us that x^3 is congruent to ± 1 modulo λ^4, as is y^3. If $x^3 + y^3$ is divisible by λ^4, then so is $\lambda^3 z^3$. But then z is divisible by λ, contrary to assumption.

Thus we have $x^3 + y^3$ congruent to ± 2 modulo λ^4, and so $\epsilon \lambda^3 z^3$ is congruent to ± 2 modulo λ^4. In particular λ divides 2. But λ has norm 3, a contradiction. ∎

Finally we do the descent step. Notice the similarity with the polynomial case.

Lemma. *If the result is true for some $m \geq 2$, then the result is true for $m - 1$.*

Sketch of proof. We rewrite the equation as

$$-\epsilon \lambda^{3m} z^3 = x^3 + y^3$$

and then we factor the right-hand side:

$$-\epsilon \lambda^{3m} z^3 = (x + y)(x + \omega y)(x + \omega^2 y).$$

Check that the difference of any two of the factors on the right is an associate of λy. Thus at most one of the factors on the right is divisible by λ^2. Since $m \geq 2$, one and hence *exactly one* of the factors on the right is divisible by λ^2 and each of the others is divisible by λ. Without loss we may assume that $x + y$ is the one that is divisible by λ^2 (and hence by λ^{3m-2}). We set

$$x + y = \lambda^{3m-2} k_0, \quad x + \omega y = \lambda k_1, \quad x + \omega^2 y = \lambda k_2.$$

Then we get

$$-\epsilon z^3 = k_0 k_1 k_2$$

and the k_i's have no common factor.

Exercise

15A.7. Justify the statement that the k_i's have no common factor. ∎

From the displayed equation and the Unique Factorization Theorem in $\mathbf{Z}[\omega]$, we see that each k_i is the associate of a cube in $\mathbf{Z}[\omega]$: $k_i = \epsilon_i u_i^3$. Now we use the equation

$$0 = 1 + \omega + \omega^2,$$

multiply both sides by $x + y$, distribute, substitute for k_i, and finally divide by $\epsilon_2 \lambda$ to get

$$u_1^3 + \epsilon_4 u_2^3 + \epsilon_5 \lambda^{3(m-1)} u_0^3 = 0,$$

with ϵ_4 and ϵ_5 units.

This is almost exactly what we wanted: an equation of the same form as the original, except with m replaced by $m - 1$. The only problem is the unit ϵ_4. If it is -1, it can be absorbed into u_2. So it remains to show

$$\epsilon_4 = \pm 1.$$

Since $m \geq 2$, λ^3 divides $u_1^3 + \epsilon u_2^3$. However λ does not divide u_1 or u_2. So as usual u_i^3 is congruent to ± 1 modulo λ^4. Hence λ^3 divides $\pm 1 \pm \epsilon_4$. Since this number has norm at most 3, it must be 0 and so

$$\epsilon_4 = \pm 1,$$

as desired. ∎

We have completed our (sketch of a) proof of FLT for $n = 3$. It is interesting to note that Wiles' proof of FLT proceeds in a completely different direction by studying the elliptic curve

$$y^2 = x(x + a^n)(x - b^n),$$

where

$$a^n + b^n = c^n$$

for some nonzero integers a, b, and c with $\gcd(a, b) = 1$.

There is a certain property of elliptic curves called **modularity**. In the 1950s Taniyama conjectured:

Every elliptic curve over **Q** is modular.

On the other hand following up on a suggestion of Gerhard Frey and work of J. P. Serre, Ken Ribet was able to prove the following in the late 1980s:

Theorem (Ribet). *If n is a prime greater than 3, then the Fermat elliptic curve $y^2 = x(x + a^n)(x - b^n)$ is not modular.*

Wiles proved the following in 1994:

Theorem (Wiles). *If A and B are distinct nonzero relatively prime integers and $AB(A - B)$ is divisible by 16, then the elliptic curve*

$$y^2 = x(x + A)(x + B)$$

is modular.

Exercise

15A.8. Explain why the theorems of Ribet and Wiles prove FLT for $n \geq 5$. ∎

Thus the old proofs of Fermat and Euler for $n = 4$ and $n = 3$ are still needed to complete the proof of FLT. A final note: Extending the work of Wiles, the full conjecture of Taniyama has now been proved by Breuil, Conrad, Diamond and Taylor.

SECTION FOUR

THE GRAND SYNTHESIS

CHAPTER 16   CONSTRUCTIBLE POLYGONS AND THE METHOD OF MR. GAUSS

CHAPTER 17   CYCLOTOMIC FIELDS AND LINEAR ALGEBRA

CHAPTER 18   A LAGRANGE THEOREM FOR FIELDS AND NONCONSTRUCTIBILITY

CHAPTER 19   GALOIS FIELDS AND THE FUNDAMENTAL THEOREM OF ALGEBRA REVISITED

CHAPTER 20   GALOIS' THEORY OF EQUATIONS

CHAPTER 21   THE GALOIS CORRESPONDENCE

CHAPTER 22   CONSTRUCTIBLE NUMBERS AND SOLVABLE EQUATIONS

CHAPTER 16

CONSTRUCTIBLE POLYGONS AND THE METHOD OF MR. GAUSS

In this unit all the players return to the stage: geometrical figures, polynomials, numbers, and groups. The first manifestation of this Grand Synthesis appears in the final section of the *Disquisitiones Arithmeticae* (1801) by Karl Friedrich Gauss. The story began five years earlier when the 19-year-old Gauss pondered the old Greek problem of constructing regular polygons with straight-edge and compass.

First let us understand what a "straight-edge and compass construction" means.

Straight-Edge and Compass Constructions. *Start with the "points"* $0 = (0, 0)$ *and* $1 = (1, 0)$. *You are allowed to*

(a) *Draw the straight line through* $0, 1$ *and/or any two points that you have "constructed."*

(b) *Set your compass at the distance between* $0, 1$ *and/or any two points that you have "constructed" and then draw the circle of that radius centered at* $0, 1$, *or any point that you have "constructed."*

A point is said to have been constructed if it is a point of intersection of two lines you have drawn or of a line and a circle that you have drawn or of two circles that you have drawn. We shall call a complex number $\alpha = a + bi$ **constructible** if the corresponding point (a, b) can be constructed by straight-edge and compass in the sense given here starting from the given points 0 and 1. In particular, a real number a is constructible if the point $(a, 0)$ is constructible.

The following exercises should be a review for you of some basic constructions from Euclidean geometry.

Exercises

16.1. Describe how to do the following basic constructions:
   a. Given points P and Q, construct the perpendicular bisector of the line segment $\overline{PQ}$.
   b. Given an angle (i.e., two rays PQ and PR emanating from a point P), construct the angle bisector.
   c. Given a line QR and a point P not on the line, construct the line through P parallel to QR.

16.2. Prove: A complex number $\alpha = a + bi$ is constructible if and only if the real numbers a and b are constructible.

16.3. **a.** Prove by induction that every integer n is constructible.
   b. Prove: Every rational number p/q is constructible. (*Hint*: One good way is to begin with the points 0, 1, p, and q, the line through 0 and 1, and one other line passing through 0. Now use similar triangles. Describe your method carefully according to the rules given earlier. In particular, remember that you can only draw a line through two points that have already been constructed and you can only draw a circle whose radius length has already been constructed. You don't have to redescribe the basic constructions from 16.1.)
   c. Illustrate your general method with one specific construction; e.g., construct 2/3.

16.4. **a.** Suppose that a is a constructible positive real number. Construct a circle of diameter $a + 1$. Construct a perpendicular to the diameter dividing it into segments of length a and 1. Prove that the half-chord from the circle to the diameter has length $\sqrt{a}$.
   b. Describe how to construct $\sqrt{\alpha}$ (with straight-edge and compass) if α is any already constructed complex number. How many square roots does α have?

16.5. Draw the circle S^1 of radius 1 centered at 0. For each of the following values of n, construct (in the sense of this chapter) the regular n-gon whose vertices lie on S^1 and one of whose vertices is 1; i.e., construct the n complex roots of $x^n - 1$ for

$$n = 2, 3, 4, 6, 8, 12.$$

Label each vertex with its corresponding (constructible) complex number written in the form $a + bi$ and also in polar form.

(*Note*: You should be able to construct the equilateral triangle by drawing at most four lines and two circles (including the circle S^1 and the line through 0 and 1).)

16.6. Prove that the triangle you constructed in 16.5 is actually equilateral. (You can certainly do this by algebra. For fun you might also want to do it in the spirit of Euclid.) ∎

OK. We know how to construct the regular n-gon for $n = 3, 4, 6, 8,$ and 12. What about 5? The regular pentagon construction is already in Euclid, but let's think about it in the manner of Gauss.

Let $\zeta = e^{2\pi i/5}$. The vertices of the regular pentagon are $1, \zeta, \zeta^2, \zeta^3$, and ζ^4. Notice that complex conjugation interchanges ζ and ζ^4 as well as ζ^2 and ζ^3. Thus the "Gauss sums" $g = \zeta + \zeta^4$ and $h = \zeta^2 + \zeta^3$ are fixed under complex conjugation and hence are real numbers. But even more is true. The four numbers ζ^i, $1 \leq i \leq 4$, are the roots of the cyclotomic polynomial

$$\Phi_5(x) = \frac{x^5 - 1}{x - 1} = x^4 + x^3 + x^2 + x + 1.$$

So

$$g + h = \zeta + \zeta^4 + \zeta^2 + \zeta^3 = -1.$$

What about gh? Well we can easily multiply it out and get

$$gh = (\zeta + \zeta^4)(\zeta^2 + \zeta^3) = \zeta^3 + \zeta^4 + \zeta + \zeta^2 = -1.$$

But this says that

$$(x - g)(x - h) = x^2 - (g + h)x + gh = x^2 + x - 1$$

and so g and h are roots of a quadratic polynomial with integer coefficients!

Exercises

16.7. Find g and h.

16.8. Show that g and h are constructible numbers.

16.9. Show that the regular pentagon is constructible.

16.10. Show that if b and c are constructible (complex) numbers, then the roots of $x^2 + bx + c$ are constructible complex numbers. ∎

Moving right along, let's consider the regular 7-gon (heptagon). Now let $\eta = e^{2\pi i/7}$. Again the six complex vertices are paired up by complex conjugation. Let $a = \eta + \eta^6$, $b = \eta^2 + \eta^5$, and $c = \eta^3 + \eta^4$. Again we have

$$a + b + c = -1.$$

Exercises

16.11. Compute abc and $ab + ac + bc$.

16.12. Find the cubic polynomial whose roots are a, b, and c.

16.13. Show that this cubic polynomial is irreducible over **Q**.

16.14. Let F be a field of constructible numbers. We say that the number α is obtained by an **elementary construction** from F if either
  (i) α is the intersection of a circle C with radius $r \in F$ and center $\beta \in F$ with a line L passing through two points γ and δ in F; or
  (ii) α is the intersection of two circles C and C' with radii r and r' in F and centers β and β' in F.

Prove: If $\alpha = a + bi$ is obtained by an elementary construction from F, then both a and b are roots of quadratic equations with coefficients in F.

16.15. Prove that if α is a real root of a cubic polynomial that is irreducible over $\mathbf{Q}$, then α cannot be obtained by an elementary construction from $\mathbf{Q}$. ∎

This does not quite prove that the regular heptagon is not constructible. Later, in Chapter 18, we shall prove this. But anyway it strongly leads one to believe that it is not constructible.

What about the regular nonagon (9-gon)? If it is constructible, then so is the regular 18-gon, and so a 20° angle is constructible and so it is possible to trisect a 60° angle by straight-edge and compass construction. But Viete's work showed a close connection between angle trisection and solvability of cubic equations.

Exercise

16.16. Using the addition formulas for cosine, find a cubic equation one of whose roots is $\cos 20°$. Prove that this cubic equation is irreducible over $\mathbf{Q}$. ∎

Now that we are in the swing of things, you may be inclined to believe that the regular n-gon is not constructible for $n = 7, 9, 11, 13,$ and 15. Actually this is true except for 15.

Exercise

16.17. Using the equilateral triangle and the regular pentagon, show how to construct the regular 15-gon. ∎

Indeed we can prove the following result.

Theorem. *If the regular m-gon and the regular n-gon are constructible and if $\gcd(m, n) = 1$, then the regular mn-gon is constructible.*

Proof. By Euclid's Lemma there exist integers a and b with

$$am + bn = 1.$$

Since $m > 1$ and $n > 1$, exactly one of a and b is positive and the other is negative. Renaming m and n if necessary and replacing b by $-b$, we may assume that $a > 0, b > 0$, and the equation is

$$am - bn = 1.$$

Multiply this equation by 2π and divide by mn to get

(16A) $$a \cdot \frac{2\pi}{n} - b \cdot \frac{2\pi}{m} = \frac{2\pi}{mn}.$$

Now since the regular n-gon and the regular m-gon are constructible, angles of $a \cdot \frac{2\pi}{n}$ radian measure and $b \cdot \frac{2\pi}{m}$ radian measure may be constructed, both having initial side as the ray through 0 and 1. The equation (16A) has the geometrical interpretation that the angle formed by the terminal sides of these angles has radian measure $\frac{2\pi}{mn}$. Since this angle is constructible, so is the regular mn-gon. ∎

Exercise

16.18. For $1 \leq n \leq 16$, which regular n-gons do we know to be constructible? Which do we think are probably not constructible? ∎

Now we have some idea why Gauss started looking at the regular 17-gon. We are now ready to begin our discussion of his work on the regular 17-gon.

In order to get constructible numbers we would like to produce quadratic equations with constructible coefficients. Let's set $\eta = e^{2\pi i/17}$. Then η and its powers are the vertices of the regular 17-gon and η is a root of the cyclotomic polynomial $\Phi_{17}(x)$ of degree 16.

We would like to break up the set of 16 roots of $\Phi_{17}(x)$ into two sets of size 8 and form the corresponding Gauss sums g and h. Then as usual $g + h = -1$. If gh is a rational number, then g and h are the two roots of a quadratic equation with rational coefficients and so they are constructible.

Then we would like to break up each set of size 8 into two sets of size 4 and form the Gauss sums a, b, c and d. If these are listed correctly, then $a + b = g$ and $c + d = h$. Again the problem is the products, but if they are OK, then a and b are the two roots of a quadratic equation with constructible (although not necessarily rational) coefficients. Likewise for c and d. So then a, b, c, and d will be constructible.

Two more such subdivisions will finally get us down to the roots of $\Phi_{17}(x)$ themselves, which will then be constructible as well.

This is Gauss' strategy. The philosophy is that 17 should be good, just as 5 was good, because 16 is a power of 2 (just as 4 is) and so we can arrive at the roots by a sequence of halvings.

But how do we get the products to be "good"? Trial and error will convince you that a random subdivision will not give a good product. We (or Gauss) have to be more clever.

Gauss noticed that among all the permutations of the roots of the equation $x^p - 1$, there are some particularly nice and important ones. Let's call them g_a and let

$$g_a(\zeta^k) = (\zeta^k)^a = \zeta^{ka}, \quad 1 \leq a \leq p - 1$$

and

$$G_p = \{g_a : 1 \leq a \leq p - 1\}$$

(g and G for Gauss).

A few things are worth noticing right away:

1. g_a is completely determined by what it does to ζ.
2. If we defined g_p in the obvious way, then it wouldn't be a permutation. In fact it would map each ζ^k to 1.

    So wait a minute. Is g_a a permutation for $1 \leq a \leq p - 1$? Yes, precisely by the argument from Euler's proof of Fermat's Little Theorem!! Why?? Because we really should think of the multiplication in the exponent ka as multiplication modulo p and then the statement that g_a is a one-to-one map is precisely (12B) of Euler from Chapter 12. (This is why I used the letter a). Consider the set

$$a, 2a, 3a, \ldots, (p-1)a.$$

Since $gcd(a, p) = 1$, no two of the numbers ia and ja are congruent modulo p unless $i = j$.

So each g_a is a permutation of the roots, by Euler's argument. Now what is the rule of composition?

3. $g_a \circ g_b = g_{ab}$, where the product ab is understood as multiplication modulo p, because

$$g_a \circ g_b(\zeta) = g_a(\zeta^b) = \zeta^{ba} = \zeta^{ab} = g_{ab}(\zeta).$$

So we have an exact correspondence between the group of permutations

$$G_p = \{g_1, g_2, \ldots, g_{p-1}\}$$

with the operation of composition of permutations and the group of numbers

$$U_p = \{1, 2, \ldots, p-1\}$$

with the operation of multiplication modulo p.

Exercise

16.19. Prove: The function $\phi : G_p \to U_p$ defined by $\phi(g_a) = a$ is an isomorphism of groups. *Note*: You have to verify that ϕ is one-to-one and also that the multiplication condition holds where $\cdot_p$ denotes multiplication modulo p):

$$\phi(g_a \circ g_b) = a \cdot_p b \text{ for all } g_a, g_b \in G_p.$$ ∎

In particular when you have two isomorphic objects, any true statement about one of them translates into a true statement about the other one. It's just like translating from one language to another. We proved in Chapter 12 that U_p is a cyclic group for every prime p. Hence we conclude that likewise the Gauss group G_p is a cyclic group for every prime p.

Specifically, when $p = 17$ and $\eta = e^{2\pi i/17}$, we can find a Gauss map,

$$g_a(\eta) = \eta^a,$$

that, when written as a permutation of the 16 roots of $\Phi_{17}(x)$, is a single cycle of length 16. Let's find this map by trial and error.

Exercises

16.20. Compute the cycle decomposition of the Gauss map g_2. Do the same for g_3. Keep going until you find a Gauss map that is a single cycle of length 16. Let's call it g. Prove that the set of all 16 Gauss maps is a cyclic group generated by the Gauss map g. Compute the cycle decompositions of the Gauss maps g^2, g^4, and g^8. Guess what the "good" Gauss sums are.

16.21. Now explicitly compute the products (and sums) for the good Gauss sums and find the sequence of quadratic equations with constructible coefficients that establish the constructibility of the regular 17-gon. Explain your work carefully.

16.22 (Bonus: For the truly stout-hearted) Develop an algorithm for constructing the regular 17-gon with straight-edge and compass. ∎

You may wonder why the regular 9-gon is not constructible, inasmuch as $9 - 1 = 8$ and 8 is a power of 2.

Exercises

16.23. Think about why this method is not applicable to the regular 9-gon. Discuss your conclusions. ∎

In the vein of the last exercise, let's consider whether we can define Ga when n is not a prime.

16.24. Let $\zeta = e^{2\pi i/n}$ be a primitive nth root of 1 and let Z_n be the set of all nth roots of 1. For a a natural number between 1 and n with $gcd(a, n) = 1$, define the Gauss map g_a on the set Z_n by

$$g_a(\zeta^k) = (\zeta^k)^a.$$

  a. Prove: g_a is a permutation of the set Z_n.
  b. Prove: The set $G_n = \{g_a : gcd(a, n) = 1, 1 \leq a \leq n\}$ is a group of permutations and $|G_n| = \varphi(n)$.

16.25. Prove: The map $f_n : U_n \to G_n$ by $f_n(a) = g_a$ is an isomorphism of groups.

16.26. Determine for which n between 2 and 24 the group U_n is a cyclic group. Can you formulate a general conjecture?

16.27. Let G be a group. We call a function $f : G \to G$ an **automorphism** of G if f is an isomorphism of the group G onto itself. Prove: If $A(G)$ is the set of all automorphisms of the group G, then $A(G)$ is itself a group under composition of functions.

16.28. Recall that the set Z_n of complex nth roots of 1 is a group under complex multiplication. Prove that G_n is the automorphism group of Z_n.

16.29. Let r be an odd prime.
  a. Prove: $(1 + r)^{r^2} \equiv 1 \pmod{r^3}$ but $(1 + r)^r \not\equiv 1 \pmod{r^3}$. (This is false if $r = 2$.)
  b. Prove: The Gauss map g_{1+r} is an element of order r^2 in the group G_{r^3}.
  c. Prove: $(g_{1+r})^r = g_{1+r^2}$ in G_{r^3}.
  d. Prove: If $\zeta = e^{2\pi i/r^3}$ and g is any Gauss map of order r in G_{r^3}, then $g(\zeta^r) = \zeta^r$.
  e. Prove: If C is any cyclic group of order r^3 with generator x and if α is any automorphism of C of order r, then $\alpha(x^r) = x^r$. (You have already proved this. Explain why.)
  f. Describe the automorphism group of Z_8 and confirm that the analogue of 16.29d is false when $r = 2$. ∎

CHAPTER 17

CYCLOTOMIC FIELDS AND LINEAR ALGEBRA

In the work of Lagrange, we blithely extended the permutation action of the symmetric group S_n acting on the roots of a polynomial $x^n + \cdots$ to a permutation action of S_n on the set of all polynomials in those roots; e.g., for $n = 3$, if r, s, and t were the roots, then we could let the permutation $\sigma = (r, s, t)$ act on polynomials in r, s, and t by permutation of the "variables." For example

$$\sigma(r + s^2) = s + t^2.$$

We run into trouble if we try to proceed in this manner with the roots, for instance, of the polynomial $x^3 - 1$. Letting $\omega = e^{2\pi i/3}$ as usual, suppose we consider the permutation

$$\sigma = (1, \omega, \omega^2).$$

This is perfectly well defined as a permutation of the roots, but if we try to extend it to a permutation of the "polynomials" in 1, ω, and ω^2, we run into an embarrassing ambiguity. What is $\sigma(\omega^4)$? Proceeding as before we say

$$\sigma(\omega^4) = (\sigma(\omega))^4 = (\omega^2)^4 = \omega^8 = \omega^2.$$

But, someone objects, $\omega^4 = (\omega^2)^2$ and so

$$\sigma(\omega^4) = (\sigma(\omega^2))^2 = 1^2 = 1.$$

Oops. There is no consistent way of extending σ to a permutation of the "polynomials" in 1, ω, and ω^2.

What happened? Well, Lagrange treated his roots as "unknowns" or independent variables. He ignored the possibility of algebraic relations among the roots. So any two polynomials in the roots that looked different were different. This is a perfectly justifiable approach to the theory of the "generic polynomial," which is what Lagrange was studying. But the cyclotomic polynomials are far from generic.

Gauss knew what the roots of his polynomials were (at least as trigonometric expressions), and there are algebraic relations among these roots that can't be ignored. Suddenly not every permutation of the roots extends in a meaningful way to a function on the polynomials in the roots.

But Gauss is only interested in certain functions—the Gauss maps. Do they extend in a meaningful way? The answer is yes. We shall take a small scenic detour via linear algebra in our attempt to confirm this. We begin by recalling the definition of a vector space.

Definition. A set V is a vector space over the field F if the following conditions hold:

There is an operation $+$ on V, called vector addition, such that $(V, +)$ is an abelian group (i.e., vector addition satisfies the associative and commutative laws, there is a zero vector, and for each vector v, there is a vector $-v$ with $v + (-v) = \mathbf{0}$.)

There is an operation $\cdot : F \times V \to V$, called scalar multiplication, satisfying the following conditions (we will write scalar multiplication simply by juxtaposition of symbols):

$1 \cdot v = v$ for all $v \in V$;

$(ab)v = a(bv)$ for $a, b \in F$ and $v \in V$;

$(a + b)v = av + bv$ for $a, b \in F$ and $v \in V$; and

$a(u + v) = au + av$ for $a \in F$ and $u, v \in V$.

We call V an F-vector space if we wish to emphasize that the field of scalars is F.

Exercises

17.1. Prove: $0 \cdot v = \mathbf{0}$ for all $v \in V$. (*Note*: 0 on the left is the scalar 0, while $\mathbf{0}$ on the right is the zero vector.)

17.2. Prove: If $a \in F - \{0\}$, then the function $L_a : V \to V$ defined by

$$L_a(v) = av$$

for all $v \in V$, is an isomorphism of the group V with itself, i.e., an automorphism of V. What is the inverse automorphism? ∎

The next observation is important for the study of fields.

17.3. Prove: If E is a field and F is any subfield of E, then E is a vector space over F. (Often we look upward from F and call E an **extension field** of F. Thus for example $\mathbf{C}$ is a vector space over $\mathbf{Q}$ and also is a vector space over $\mathbf{R}$. Likewise $\mathbf{R}$ is a vector space over $\mathbf{Q}$.)

Definition. A subset U of the vector space V over F is called a **subspace** of V if $(U, +)$ is a subgroup of $(V, +)$ (i.e., $(U, +)$ is closed under vector addition and subtraction) and U is closed under scalar multiplication (i.e., if $c \in F$ and $u \in U$, then $cu \in U$).

17.4. Prove: Every subfield of $\mathbf{C}$ is a $\mathbf{Q}$-subspace of the $\mathbf{Q}$-vector space $\mathbf{C}$. ∎

Now although elementary, this vector-space point of view is very powerful, in particular because of the concept of a basis for a vector space. Since we shall only be dealing with finite-dimensional vector spaces, this will provide a finite number to "measure" our vector spaces, which have infinite cardinality as sets but which as we shall see behave in a very "finite" way. Let's quickly develop the basic facts about finite-dimensional vector spaces over F.

Definition. Let V be a vector space. A subset S of V is called a **spanning set** for V if every vector in V can be written as a finite linear combination of vectors in S; i.e., if $v \in V$, there exist $s_1, \ldots, s_n \in S$ and scalars $a_1, \ldots, a_n \in F$ with

$$v = a_1 s_1 + \cdots + a_n s_n.$$

S is called a **basis** for V if this expression is unique for each $v \in V$.

Theorem. *If V is a vector space over F with a finite spanning set S, then S contains a subset B that is a basis for V over F.*

Proof. Let B be a subset of S that is minimal with respect to being a spanning set for V. Suppose some vector v has two expressions as a linear combination of vectors from B:

$$a_1 s_1 + \cdots + a_n s_n = v = c_1 s_1 + \cdots + c_n s_n.$$

Without loss we may assume that $a_1 \neq c_1$. But then

$$(a_1 - c_1)s_1 = (c_2 - a_2)s_2 + \cdots + (c_n - a_n)s_n$$

and since F is a field and $a_1 - c_1 \neq 0$, we may multiply both sides by $(a_1 - c_1)^{-1}$ to get s_1 expressed as a linear combination of $s_2, \ldots, s_n$. But then $B - \{s_1\}$ is a smaller spanning set for V, contrary to the choice of B, a contradiction. Hence B is a basis for V. ∎

In order to check uniqueness we only have to look at the zero vector. This is the source of the concept of linear independence.

Definition. A set S of vectors is **linearly independent** if the zero vector is uniquely expressible as a linear combination of vectors from S. Otherwise S is said to be linearly dependent.

Theorem. *If V is a vector space over F and S is a linearly independent spanning set for V, then S is a basis for V.*

Proof. In the previous proof, bring the right-hand side of the first equation over to the left-hand side:

$$(a_1 - c_1)s_1 + \cdots + (a_n - c_n)s_n = \mathbf{0}.$$

Since $\mathbf{0}$ is uniquely expressible as a linear combination of vectors from S, we must have $a_i = c_i$ for all i, as desired. ∎

Once we have a finite basis $B = \{s_1, \ldots, s_n\}$ for V, we have an unambiguous way of assigning an "address" to each vector in V; namely we have an isomorphism

$$\Phi : V \to F^n$$

defined by
$$\Phi(v) = (a_1, \ldots, a_n) \in F^n$$
if
$$v = a_1 s_1 + \cdots + a_n s_n.$$

Clearly this is a one-to-one mapping of the vector space V onto the vector space F^n satisfying
$$\Phi(u + v) = \Phi(u) + \Phi(v)$$
and
$$\Phi(cu) = c\Phi(u)$$
for all vectors u and v and scalars c. This constitutes the definition of an **isomorphism of vector spaces**. Thus we have the following theorem:

Theorem. *Every nonzero vector space V over the field F with a finite spanning set is isomorphic to F^n for some positive integer n.*

Our next goal is to show that n is unique. We shall then call n the **dimension** of the vector space V over the field F.

Theorem. *If $F^n \cong F^m$, then $n = m$.*

Proof. Suppose $F^n \cong F^m$ with $n \geq m$. Then F^m has a basis $B = \{v_1, \ldots, v_n\}$ of cardinality n. Write
$$v_i = (c_{1i}, \ldots, c_{mi}) \in F^m$$
for $1 \leq i \leq n$. The statement that B is a linearly independent set is equivalent to the statement that the system of m homogeneous equations in n unknowns:
$$c_{11}x_1 + \cdots + c_{1n}x_n = 0$$
$$\cdots$$
$$c_{m1}x_1 + \cdots + c_{mn}x_n = 0$$
has only the trivial solution $x_1 = \cdots = x_n = 0$. Gaussian elimination on the other hand shows that such a system has infinitely many solutions when $n > m$. Hence $n = m$, as claimed. ∎

Not all vector spaces have a finite basis. For instance **C** is not a finite-dimensional vector space over **Q**. On the other hand **C** is a two-dimensional vector space over **R**.

An important class of examples of finite-dimensional vector spaces will be afforded by the splitting fields for polynomial equations.

Definition. Let F be a subfield of **C** and let $p(x)$ be a polynomial in $F[x]$. An extension field E of F is called the **splitting field** for $p(x)$ if E is the smallest extension field of F containing all of the roots of $p(x)$. Sometimes we may refer to extension fields of E as splitting fields for $p(x)$ as well.

Definition. Let F be a subfield of $\mathbf{C}$ and let α be any complex number. Then $F(\alpha)$ is the smallest extension field of F (inside of $\mathbf{C}$) that contains α.

You might well wonder if $F(\alpha)$ is well defined. Is there really a unique smallest subfield of $\mathbf{C}$ containing F and α? This is answered by the following theorem.

Theorem. *Let $\{E_i\}_{i \in I}$ be a collection of subfields of $\mathbf{C}$. Then the intersection E of all of the E_i's is again a subfield of $\mathbf{C}$. In particular if F is any subfield of $\mathbf{C}$ and α is any number in $\mathbf{C}$, then $F(\alpha)$ is the intersection of all extension fields of F containing α.*

Proof. Since every E_i contains 0 and 1, so does E. (Indeed E contains $\mathbf{Q}$.) Now if r and s are in E, then for any i, we have that r and s are in E_i and so $r + s, -s, rs$, and (if $s \neq 0$) r/s are in E_i for all i, hence in E. Thus E is indeed a subfield of $\mathbf{C}$. Now clearly the intersection of all extension fields of F containing α is the unique smallest extension field of F containing α. ∎

Now let F be any field and let $p(x) = x^2 + bx + c = 0$ be a quadratic equation with $b, c \in F$. Let r and s be the two roots of $p(x)$ and let $\delta = \sqrt{b^2 - 4c}$.

Exercises

17.5. Prove: $F(r) = F(s) = F(\delta) = \{a + dr : a, d \in F\} = \{a + d\delta : a, d \in F\}$.
(*Hint*: It is easiest to show that $\{a + d\delta : a, d \in F\}$ is a field. *Caution*: Be sure that you are not dividing by 0. The case when $\delta \in F$ is an easy special case.)

17.6. Prove: $F(r)$ is one- or two-dimensional as a vector space over F. ∎

Notice that $F(r)$ is a splitting field for the polynomial $p(x)$. Conversely we have the following result.

Theorem. *Let F and E be subfields of $\mathbf{C}$ with E an extension field of F. Suppose that E is two-dimensional as an F-vector space. Then $E = F(r)$ is the splitting field of an irreducible quadratic polynomial with coefficients in F.*

Proof. Let $\{1, r\}$ be a basis for K as an F-vector space. Since K is a field, $r^2 \in K$ and so $r^2 = c1 + br$ for some $b, c \in F$. Thus r is a root of the quadratic polynomial $p(x) = x^2 - bx - c$ with coefficients in F. If $p(x)$ is not irreducible, then $r \in F$. But then $\{1, r\}$ is a linearly dependent set, contrary to assumption. ∎

Life is never this easy again. Unfortunately (?) the general situation is quite a bit more complicated and complications start as soon as we move from quadratic to cubic equations. For example, let r be the real root of the cubic polynomial $x^3 - 2$ and let $\mathbf{Q}(r)$ be the smallest field containing $\mathbf{Q}$ and r.

First problem: We can't expect to have $\mathbf{Q}(r) = \{a + br : a, b \in \mathbf{Q}\}$ because then as earlier, r would satisfy a quadratic polynomial over $\mathbf{Q}$. Well, that's not so hard to fix. We can take $\mathbf{Q}(r) = \{a + br + cr^2 : a, b, c \in \mathbf{Q}\}$, and with a little effort we can show that this is a field.

Exercises

Throughout let r be the real cube root of 2.

17.7. Find formulas for the sum

$$(a + br + cr^2) + (d + er + fr^2)$$

and the product

$$(a + br + cr^2)(d + er + fr^2)$$

taking into account the fact that $r^3 = 2$.

17.8. Compute explicitly $1/(1 + 2r)$ as a polynomial in r.

17.9. Find a general formula for $1/(a + br)$, $b \neq 0$; i.e., find formulas for x, y, and z satisfying

$$(a + br)(x + yr + zr^2) = 1.$$

Confirm that this is always valid, i.e., that the denominators are never 0.

17.10. Compute explicitly $1/(1 + 2r - r^2)$ as a polynomial in r.

17.11. Find a general formula for $1/(a + br + cr^2)$, $c \neq 0$. Again confirm that the denominators are never 0.

17.12. Think about the following: In deriving the formulas for addition, subtraction, multiplication, and division in $\mathbf{Q}(r)$, what facts about r did you use? Suppose that s is a nonreal cube root of 2, what would the corresponding formulas in $\mathbf{Q}(s)$ look like? What does this say about $\mathbf{Q}(r)$ and $\mathbf{Q}(s)$? ■

Now let $a \in \mathbf{C}$ be a root of a polynomial $f(x) \in \mathbf{Q}[x]$. Define

$$\mathbf{Q}(a) = \{p(a) : p(x) \in \mathbf{Q}[x]\}.$$

We first introduce the minimum polynomial of a over $\mathbf{Q}$.

Theorem. *Let F be a subfield of $\mathbf{C}$ and let $a \in \mathbf{C}$ be a root of some nonzero polynomial with coefficients in F. Then there is a unique monic polynomial $m(x) \in F[x]$ of smallest degree having a as a root. Moreover $m(x)$ is characterized as the unique monic irreducible polynomial in $F[x]$ having a as a root. If $p(x)$ is any polynomial with $p(a) = 0$, then $p(x)$ is a multiple of $m(x)$.*

Proof. First note that since a is the root of some nonzero polynomial in $F[x]$, we may choose a monic polynomial $m(x)$ of minimum degree with $m(a) = 0$ (by the Well-Ordering Principle for the positive integers). If $m(x)$ is not irreducible, then we have

$$m(x) = f(x)g(x)$$

for some $f(x)$ and $g(x) \in F[x]$ with $deg(f) < deg(m)$ and $deg(g) < deg(m)$. But then $f(a) \neq 0$ and $g(a) \neq 0$ by minimal choice of $deg(m)$. On the other hand

$$m(a) = 0 = f(a)g(a),$$

a contradiction, since F is a field. Hence a is a root of the monic irreducible polynomial $m(x)$.

Now let $p(x)$ be any polynomial in $F[x]$ with a as a root. By the Division Algorithm, there exist polynomials $q(x)$ and $r(x) \in F[x]$ with

$$p(x) = q(x)m(x) + r(x)$$

and with $r(x) = 0$ or $deg(r) < deg(m)$. Plug in a:

$$0 = p(a) = q(a)m(a) + r(a) = 0 + r(a) = r(a).$$

Since $r(a) = 0$, minimal choice of $m(x)$ forces $r(x) = 0$. Thus $p(x)$ is a multiple of $m(x)$. In particular $m(x)$ is the unique monic irreducible polynomial having a as a root. ∎

Theorem. $\mathbf{Q}(a)$ *is a subfield of* $\mathbf{C}$. *Indeed it is the smallest subfield of* $\mathbf{C}$ *containing a.*

Proof. Clearly $\mathbf{Q}(a)$ is closed under addition, subtraction, and multiplication. Also clearly it is the smallest subset of $\mathbf{C}$ containing a and 1 and closed under these operations. So it will suffice to prove that $\mathbf{Q}(a)$ is closed under division (by nonzero elements). It will suffice to prove the following:

If $g(x) \in \mathbf{Q}[x]$ with $g(a) \neq 0$, then there exists a polynomial $f(x) \in \mathbf{Q}[x]$ with

$$\frac{1}{g(a)} = f(a);$$

i.e.,

$$f(a)g(a) = 1.$$

Let $m(x) \in \mathbf{Q}[x]$ be the minimum polynomial for a over $\mathbf{Q}$. If $g(x)$ is a multiple of $m(x)$, then $g(a) = 0$, contrary to assumption. But then since $m(x)$ is irreducible in $\mathbf{Q}[x]$, we must have that $gcd(m(x), g(x)) = 1$. Then Euclid's Lemma implies that there exist polynomials $f(x), h(x) \in \mathbf{Q}[x]$ such that

$$f(x)g(x) + h(x)m(x) = 1.$$

Now plug in a:

$$f(a)g(a) + h(a)m(a) = f(a)g(a) + 0 = f(a)g(a) = 1,$$

as desired. ∎

Thus $\mathbf{Q}(a)$ is indeed a field. There is a more elementary but somewhat more complicated proof of this that was discovered by my student, Jason Petry. We present this proof in Appendix B to this section.

We shall pursue the subject of splitting fields further in the next chapter. Right now we ease our way back toward Gauss and the cyclotomic splitting fields, which can in fact be constructed by a single extension. Specifically let's again consider the 17th roots of 1. Again let $\eta = e^{2\pi i/17}$ and consider

$$\mathbf{Q}(\eta) = \{a_0 + a_1\eta + a_2\eta^2 + \cdots + a_{16}\eta^{16} : a_0, a_1, \ldots, a_{16} \in \mathbf{Q}\}.$$

Clearly $\mathbf{Q}(\eta)$ is a $\mathbf{Q}$-vector subspace of $\mathbf{C}$ of dimension at most 17 (since the set $\{1, \eta, \eta^2, \ldots, \eta^{16}\}$ has cardinality 17). In fact since

$$1 + \eta + \eta^2 + \cdots + \eta^{16} = 0,$$

we see that $\{\eta, \eta^2, \ldots, \eta^{16}\}$ is a spanning set for $\mathbf{Q}(\eta)$.

Moreover since $\eta^n = \eta^r$ if $n = 17q + r$ with $0 \le r \le 16$, we see that
$$\mathbf{Q}(\eta) = \{p(\eta) : p(x) \in \mathbf{Q}[x]\}.$$
Thus $\mathbf{Q}(\eta)$ is the splitting field for the polynomial $x^{17} - 1$.

Suppose that $\{\eta, \eta^2, \ldots, \eta^{16}\}$ is a basis for $\mathbf{Q}(\eta)$. Then the Gauss maps are extendible to the set of all polynomials in η (i.e., to the field $\mathbf{Q}(\eta)$) by the following general principle:

Definition. Let V be a vector space over F. A linear transformation (or linear operator) on V is a function $T : V \to V$ satisfying

1. $T(u + v) = T(u) + T(v)$ for all $u, v \in V$; and
2. $T(cv) = cT(v)$ for all $c \in F$ and $v \in V$.

Theorem. *Let V be a vector space over F with basis B. Then any function $f : B \to V$ is uniquely extendible to a linear transformation $T : V \to V$.*

Proof. For any $v \in V$, write v uniquely as
$$v = a_1 s_1 + \cdots + a_n s_n$$
with $a_i \in F$ and $s_i \in B$ for all i. Define
$$T(v) = a_1 f(s_1) + \cdots + a_n f(s_n).$$
Then T is well defined by the uniqueness of the expression for v. Clearly T extends f, and it is easy to check that T is a linear transformation. ∎

But first we need to know that $\{\eta, \eta^2, \ldots, \eta^{16}\}$ is a basis for $\mathbf{Q}(\eta)$. This turns out to be intimately connected to the irreducibility of the cyclotomic polynomial $\Phi_{17}(x)$, as Gauss was well aware. (Gauss was not in fact familiar with the language of vector spaces. The terminology of scalars and vectors was first introduced around 1850 by William Rowan Hamilton in the context of quaternions.)

Exercise

17.13. Prove: If $\Phi_{17}(x)$ is an irreducible polynomial, then $\{\eta, \eta^2, \ldots, \eta^{16}\}$ is a linearly independent set and hence a basis for $\mathbf{Q}(\eta)$ as a vector space over $\mathbf{Q}$. ∎

All roads to the irreducibility of $\Phi_{17}(x)$ seem to lead through Gauss' Lemma.

Definition. Let
$$f(x) = a_0 + a_1 x + \cdots + a_n x^n \in \mathbf{Z}[x].$$
We say that the **content** of f, $c(f)$, is the (positive) greatest common divisor of the coefficients $a_0, a_1, \ldots, a_n$.

Lemma. *If f and g are polynomials in $\mathbf{Z}[x]$, then*
$$c(fg) = c(f)c(g).$$

Proof. Clearly $c(f)c(g)$ divides every coefficient of $f(x)g(x)$ and so dividing through by $c(f)c(g)$, we may assume that $c(f) = c(g) = 1$ and try to show that $c(fg) = 1$. If not, there is a prime p that divides every coefficient of the polynomial $fg(x) = f(x)g(x)$. But then interpreting each of these polynomials as polynomials in the domain $D = \mathbf{Z}/p\mathbf{Z}[x]$, we have that $\overline{fg} = 0 \in D$ but $\overline{f} \neq 0 \neq \overline{g}$, contrary to the Cancellation Law in the domain D. ∎

Gauss' Lemma. *Let $p(x)$ be a polynomial in $\mathbf{Z}[x]$ with $c(p) = 1$. Then $p(x)$ is irreducible in $\mathbf{Z}[x]$ if and only if $p(x)$ is irreducible in $\mathbf{Q}[x]$.*

Proof. First assume that $p(x)$ is reducible in $\mathbf{Z}[x]$ and write $p(x) = f(x)g(x) \in \mathbf{Z}[x]$ with neither f nor g a constant polynomial. Then $p(x) = f(x)g(x)$ is a nontrivial factorization of $p(x)$ in $\mathbf{Q}[x]$ as well.

Next suppose that $p(x)$ can be factored in the (larger) domain $\mathbf{Q}[x]$. Write $p(x) = f(x)g(x)$ with $f(x)$ and $g(x)$ in $\mathbf{Q}[x]$ with neither a unit in $\mathbf{Q}[x]$ and hence both of lesser degree than $p(x)$. Let m_f and m_g be the *lcm*'s of the denominators of the coefficients of f and g respectively. Let $m = m_f m_g$. Then multiplying both sides by m we get

$$mp(x) = f_1(x)g_1(x)$$

with $f_1(x) = m_f f(x)$ and $g_1(x) = m_g g(x)$ in $\mathbf{Z}[x]$. By the previous lemma, we may equate the contents on both sides:

$$m = mc(p) = c(f_1)c(g_1).$$

Write $f_1(x) = c(f_1)f_2(x)$ and $g_1(x) = c(g_1)g_2(x)$ with $f_2(x)$ and $g_2(x) \in \mathbf{Z}[x]$. Then

$$mp(x) = f_1(x)g_1(x) = c(f_1)c(g_1)f_2(x)g_2(x) = mf_2(x)g_2(x)$$

and so

$$p(x) = f_2(x)g_2(x).$$

Thus $p(x)$ is reducible in $\mathbf{Z}[x]$, as desired. ∎

With Gauss' Lemma in hand we have a couple of avenues to the proof of the irreducibility of $\Phi_{17}(x)$.

Theorem (Gauss). *$\Phi_p(x)$ is irreducible in $\mathbf{Q}[x]$ for all primes p.*

Proof (Kronecker). Suppose on the contrary that $\Phi_p(x)$ is not irreducible in $\mathbf{Q}[x]$. Then by Gauss' Lemma we may write

$$\Phi_p(x) = f(x)g(x)$$

with f and g monic polynomials in $\mathbf{Z}[x]$. Since $\Phi_p(1) = p$ and p is a prime, we may assume that $f(1) = \pm 1$. Let η be a primitive pth root of 1 with $f(\eta) = 0$. Then η^k is a root of $f(x^l)$ where $kl \equiv 1 \pmod{p}$. Hence every root of $\Phi_p(x)$ is a root of the polynomial

$$P(x) = f(x)f(x^2)\cdots f(x^{p-1}).$$

Thus $\Phi_p(x)$ divides $P(x)$ in $\mathbf{Q}[x]$. It follows from 12.30e that $\Phi_p(x)$ is a monic polynomial in $\mathbf{Z}[x]$ and then by 12.30d that we have

$$P(x) = \Phi_p(x)q(x)$$

for some $q(x) \in \mathbf{Z}[x]$. But then as $f(1^k) = f(1) = \pm 1$ for all k, we have

$$P(1) = (\pm 1)^{p-1} = \Phi_p(1)q(1) = p \cdot q(1),$$

with $q(1) \in \mathbf{Z}$. This is absurd and so $\Phi_p(x)$ is irreducible in $\mathbf{Q}[x]$ as claimed. ∎

Another approach to the irreducibility of $\Phi_p(x)$ was discovered by F. G. Eisenstein.

Eisenstein's Irreducibility Criterion. *Let p be a prime and let*

$$f(x) = x^n + a_{n-1}x^{n-1} + \cdots + a_1 x + a_0 \in \mathbf{Z}[x]$$

with each a_i divisible by p but a_0 not divisible by p^2. Then $f(x)$ is irreducible in $\mathbf{Q}[x]$.

Proof. Suppose on the contrary that $f(x)$ is not irreducible in $\mathbf{Q}[x]$. By Gauss' Lemma, we can find polynomials

$$g(x) = x^m + b_{m-1}x^{m-1} + \cdots + b_1 x + b_0$$

and

$$h(x) = x^r + c_{r-1}x^{r-1} + \cdots + c_1 x + c_0,$$

with

$$f(x) = g(x)h(x),$$

with all of the b_i and c_j integers and with $m < n$ and $r < n$. Then $a_0 = b_0 c_0$ and so exactly one of b_0 and c_0 is divisible by p. We may assume that b_0 is divisible by p but c_0 is not.

As in the proof of the lemma about contents of polynomials, we may interpret the three polynomials f, g, and h as having coefficients in the field $\mathbf{Z}/p\mathbf{Z}$ of integers modulo p. If we write these polynomials as $\overline{f}, \overline{g},$ and $\overline{h}$, we have by hypothesis

$$\overline{f} = x^m.$$

We still have

$$x^m = \overline{f} = \overline{g} \cdot \overline{h}.$$

But then $\overline{g} = x^m$ and $\overline{h} = x^r$, contrary to the fact that $\overline{c}_0 \neq 0$. ∎

Exercises

17.14. Prove: If $f(x) \in F[x]$ and $a \in F$, then $f(x)$ is irreducible if and only if $f(x+a)$ is irreducible. (*Note*: This is not true for nonlinear substitutions. We can easily have $f(x)$ irreducible but $f(x^2)$ not irreducible, for instance.)

17.15. Prove: If p is a prime, then $\Phi_p(x+1)$ is irreducible in $\mathbf{Q}[x]$, whence $\Phi_p(x)$ is also irreducible in $\mathbf{Q}[x]$. (In particular $\Phi_{17}(x)$ is irreducible in $\mathbf{Q}[x]$.) (*Hint*: Apply Eisenstein's Criterion to $\Phi_p(x+1)$.)

17.16. For which values of n is $\Phi_n(x)$ irreducible in $\mathbf{R}[x]$?

17.17. Imitate Kronecker's proof to show that for every prime power p^n, the cyclotomic polynomial $\Phi_{p^n}(x)$ is irreducible in $\mathbf{Q}[x]$. ∎

We have now established the irreducibility of $\Phi_{17}(x)$ in $\mathbf{Q}[x]$. Thus

$$\{\eta, \eta^2, \ldots, \eta^{16}\}$$

is a basis for $\mathbf{Q}(\eta)$ as a vector space over $\mathbf{Q}$ and each of the Gauss maps g_a is uniquely extendible to a linear transformation

$$g_a : \mathbf{Q}(\eta) \to \mathbf{Q}(\eta).$$

Even better, these maps are multiplicative:

$$g_a(f(\eta)h(\eta)) = fh(\eta^a) = f(\eta^a)h(\eta^a) = g_a(f(\eta))g_a(h(\eta)).$$

Definition. If F and K are fields and $g : F \to K$ is a one-to-one function of F onto K, then we say that g is an **isomorphism of fields** provided that

$$g(a+b) = g(a) + g(b) \text{ and } g(ab) = g(a)g(b) \text{ for all } a, b \in F.$$

If $F = K$, we say that g is an **automorphism** of the field F.

Thus we have

Theorem. *The Gauss maps $g_a : \mathbf{Q}(\eta) \to \mathbf{Q}(\eta)$ for $1 \leq a \leq 16$ are automorphisms of the field $\mathbf{Q}(\eta)$.*

Exercise

17.18. What is the inverse of the automorphism g_2? Same question for g_3. ∎

Finally we can give an easy (Ha, Ha!) proof of the constructibility of the regular 17-gon. Recall the setup. We found a Gauss map g that was a single cycle of length 16 on the 16 primitive 17th roots of 1. Thus g^2 is a product of two orbits of size 8. Let q and r be the (Gauss) sums of the numbers in these two orbits of size 8. Let c, d, e, and f be the Gauss sums of the numbers in the four orbits of size 4 for g^4, chosen so that $c + d = q$ and $e + f = r$. Let s, t, u, v, w, x, y, and z be the Gauss sums of the numbers in the eight orbits of size 2 for g^8, chosen so that $s + t = c, u + v = d, w + x = e$, and $y + z = f$.

Exercises

17.19. Let $\alpha = \Sigma_{i=1}^{16} a_i \eta^i$ with a_i a rational number for all i. Suppose that $g(\alpha) = \alpha$ (for g as earlier). Prove that $a_i = a_1$ for all i. Conclude that α is a rational number.

17.20. Prove that $g(q + r) = q + r$ and $g(qr) = qr$ without explicitly calculating the values of $q + r$ and qr. Conclude that q and r are the roots of a quadratic polynomial with rational coefficients.

17.21. Let α be as in 17.19, except now suppose that $g^2(\alpha) = \alpha$. (Do not assume that $g(\alpha) = \alpha$.) Prove that $\alpha = aq + a'r$ with q and r the Gauss sums given earlier and a and a' rational numbers.

17.22. Prove that $g^2(cd) = cd$ and $g^2(ef) = ef$. Conclude that c and d are the two roots of a quadratic polynomial with coefficients in the field $\mathbf{Q}(q, r)$. Reach a similar conclusion for e and f. Conclude that c, d, e, and f are constructible numbers. ∎

A similar argument shows that s, t, etc. are constructible numbers and finally that η and its powers are constructible numbers. Thus the regular 17-gon is constructible.

Of course we knew this already by direct calculation, so why the big effort? Well, our new argument permits generalization.

17.23. Verify that 257 is a prime. Sketch an argument that the regular 257-gon is constructible with straight-edge and compass.

17.24. Guess for which values of n the regular n-gon is constructible with straight-edge and compass. ∎

We have seen that each Gauss map g_a is an automorphism of the field $\mathbf{Q}(\eta)$. The set G_{17} of all Gauss maps of $\mathbf{Q}(\eta)$ is a group of automorphisms of $\mathbf{Q}(\eta)$. It is what will later come to be known as the **Galois group** of $\mathbf{Q}(\eta)$ or (equivalently) the Galois group of the equation $\Phi_{17}(x)$. It is the subgroup of the group of all permutations of the roots of an equation that "remembers" the algebraic relations among the roots of that equation.

Although Gauss used this group to great effect in analyzing the cyclotomic polynomials, he did not pursue the full implications of his work. Nils Henrik Abel studied Gauss' work and saw how to extend it to a larger class of equations whose Galois groups came to be known as **abelian groups**. It remained for Evariste Galois to see clearly the true significance of what he called "*la méthode de M. Gauss.*"

APPENDIX A   THE IRREDUCIBILITY OF $\Phi_n(x)$

We now sketch Gauss' proof of the irreducibility of $\Phi_n(x)$ for all n.

Theorem. *For all positive integers n, $\Phi_n(x)$ is irreducible in $\mathbf{Q}[x]$.*

Proof. Suppose not. We claim that there is a prime p with $\gcd(p, n) = 1$ and a root η of $\Phi_n(x)$ such that η and η^p do not have the same minimum polynomial in $\mathbf{Q}[x]$. First recall that every root of $\Phi_n(x)$ is η^k for some k with $\gcd(k, n) = 1$ and conversely every such η^k is a root of $\Phi_n(x)$. Write $k = p_1 p_2 \cdots p_r$ with p_i prime. Then $\gcd(p_i, n) = 1$ for all i. If η and η^{p_1} have the same minimum polynomial, and then η^{p_1} and $(\eta^{p_1})^{p_2}$ have the same minimum polynomial, etc., then eventually η and η^k have the same minimum polynomial for all k with $\gcd(k, n) = 1$. But then this minimum polynomial must be $\Phi_n(x)$ and we are done.

Thus we may assume that η has monic minimum polynomial $g(x)$ and η^p has monic minimum polynomial $f(x)$ with $f(x)$ and $g(x)$ distinct irreducible factors of $\Phi_n(x)$ in $\mathbf{Z}[x]$. Then η is a root of the monic polynomial $f(x^p) \in \mathbf{Z}[x]$. Write

$$f(x^p) = (x^p)^m + a_1(x^p)^{m-1} + \cdots + a_m.$$

By Fermat's Little Theorem, $a_i^p \equiv a_i \pmod{p}$ for all i. Thus we have

$$f(x^p) \equiv (x^m)^p + a_1^p(x^{m-1})^p + \cdots + a_m^p \equiv (x^m + a_1 x^{m-1} + \cdots + a_m)^p$$
$$\equiv f(x)^p \pmod{p}.$$

On the other hand, as η is a root of $f(x^p)$ and $g(x)$ is the monic minimum polynomial for η in $\mathbf{Z}[x]$, we have

$$f(x^p) = g(x)k(x)$$

for some monic polynomial $k(x) \in \mathbf{Z}[x]$.

Now we pass to the domain $D = \mathbf{Z}/p\mathbf{Z}[x]$ and write $\bar{h}$ as the image in D of the polynomial $h(x) \in \mathbf{Z}[x]$. Then setting $f_1(x) = f(x^p)$, the preceding equations and congruences yield

$$\bar{f}^p = \bar{f}_1 = \bar{g}\bar{k}.$$

Thus $\gcd(\bar{f}, \bar{g}) \neq 1$. As

$$x^n - \bar{1} = \bar{f}\bar{g} \cdots \in D,$$

$x^n - \bar{1}$ has a multiple irreducible factor $\bar{h} \in D$. But then $\bar{h}$ is a factor of the derivative $\bar{n}x^{n-1}$ of $x^n - \bar{1}$ in D. Since $\bar{n} \neq \bar{0}$, it follows that $\bar{h}(x) = x$ is a factor of $x^n - \bar{1}$, which is clearly absurd.

This contradiction completes the proof. ∎

APPENDIX B   DIVISIBILITY IN $F(r)$

Although the preceding approach is the "royal road" to field extensions, it is interesting to note the following elementary approach. Let α be a root of the irreducible polynomial $p(x) = x^n + a_1 x^{n-1} + \cdots + a_n \in F[x]$. Let

$$E = \{g(\alpha) : g(x) \in F[x]\}.$$

Exercise

17.25. Prove:

$$E = \{g(\alpha) : g(x) \in F[x], \deg(g(x)) < n \text{ or } g(x) = 0\}.$$

As we have seen E is a ring of numbers containing F and α. What remains unclear is, Is E closed under division? Clearly it is enough to answer the following question:

$$\text{Is } \frac{1}{g(\alpha)} \in E \text{ for each } g(\alpha) \neq 0?$$

Let's be as naive as possible. We are hoping for a polynomial $h(x) \in F[x]$ with $\deg h(x) < n$ such that $h(\alpha) = 1/g(\alpha)$; i.e., $g(\alpha)h(\alpha) = 1$. Optimistically write

$$h(\alpha) = x_0 + x_1\alpha + x_2\alpha^2 + \cdots + x_{n-1}\alpha^{n-1}$$

where $x_0, x_1, \ldots, x_n$ are "unknown" numbers in F. Multiplying $h(\alpha)$ by $g(\alpha)$ (which has "known" coefficients in F) will lead to an equation,

$$g(\alpha)h(\alpha) = 1 = 1 + 0\alpha + 0\alpha^2 + \cdots + 0\alpha^{n-1},$$

where the left-hand side is a polynomial of degree at most $n-1$ in α whose coefficients are F-linear combinations of the x_i's. Since $\{1, \alpha, \alpha^2, \ldots, \alpha^{n-1}\}$

is a basis for E_0, we may equate coefficients. This gives a system of n linear equations in n unknowns, which we need to solve. ∎

Unfortunately it is not obvious that this system has a solution. However, here is a clever trick that I learned from my student, Jason Petry. Temporarily forget the equation arising from equating the constant terms. What is left is a homogeneous system of $n-1$ linear equations in n unknowns. By basic linear algebra, such a system always has infinitely many solutions. In particular there is certainly a solution other than the trivial solution: $x_0 = x_1 = \cdots = x_{n-1} = 0$. Choose *any* nontrivial solution:

$$h(\alpha) = c_0 + c_1\alpha + \cdots + c_{n-1}\alpha^{n-1}.$$

Then since the set $\{1, \alpha, \ldots, \alpha^{n-1}\}$ is linearly independent, $h(\alpha) \neq 0$. Since also $g(\alpha) \neq 0$ and since both $h(\alpha)$ and $g(\alpha)$ are complex numbers, we have

$$g(\alpha)h(\alpha) = c \in F \text{ with } c \neq 0.$$

But now we may set

$$h_1(\alpha) = \frac{1}{c}h(\alpha) = \frac{c_0}{c} + \frac{c_1}{c}\alpha + \cdots + \frac{c_{n-1}}{c}\alpha^{n-1} \in E_0$$

and we have what we want:

$$g(\alpha)h_1(\alpha) = \frac{c}{c} = 1.$$

Thus we have given a second proof of the following theorem.

Theorem. *Let F be any subfield of* $\mathbf{C}$. *Let α be a complex root of the irreducible polynomial $p(x)$ of degree $n \geq 1$. Then*

$$F(\alpha) = \{b_0 + b_1\alpha + \cdots + b_{n-1}\alpha^{n-1} : b_0, b_1, \ldots, b_{n-1} \in F\}$$

is the smallest subfield of $\mathbf{C}$ *containing both F and α.*

CHAPTER 18

A LAGRANGE THEOREM FOR FIELDS AND NONCONSTRUCTIBILITY

We have now mastered the art of constructing simple algebraic field extensions; i.e., of adjoining one root of a polynomial $p(x)$ to a field containing the coefficients of $p(x)$. We now turn to the general problem of constructing splitting fields.

To build splitting fields we usually have to perform this extension process several times.

For example, going back to our polynomial $x^3 - 2$ and its real root r, we notice that $\mathbf{Q}(r)$ contains only real numbers. But the other two roots of $x^3 - 2$ are nonreal complex numbers. So they can't possibly lie in the field $\mathbf{Q}(r)$. So $\mathbf{Q}(r)$ is a field containing r, but it isn't the splitting field of the polynomial $x^3 - 2$.

Exercise

18.1. Let s and t be the two nonreal roots of $x^3 - 2$. Express s and t in complex polar form. Explicitly find the quadratic minimum polynomial for s and t in $\mathbf{Q}(r)[x]$. Prove that a splitting field for $x^3 - 2$ is $\mathbf{Q}(r, \omega)$ where ω is a nonreal cube root of 1. ■

Definition. Let F be any subfield of $\mathbf{C}$ and let E be an extension field of F that is n-dimensional as an F-vector space. We say that the **degree** of E over F is n and write $(E : F) = n$. ■

Now let $E = \mathbf{Q}(r, \omega)$ be the splitting field of $x^3 - 2$ over $\mathbf{Q}$. We would like to compute $(E : \mathbf{Q})$.

Exercise

18.2. Write down a basis for $\mathbf{Q}(r)$ as a $\mathbf{Q}$-vector space. Next write down a basis for E as a $\mathbf{Q}(r)$-vector space. Now try to write down a basis for E as a $\mathbf{Q}$-vector space. ■

The following important theorem may be thought of as a Lagrange Theorem for Field Extensions. (It was, however, probably not known to Lagrange.)

A Lagrange Theorem for Field Extensions. *Let $F \subseteq K \subseteq E \subseteq \mathbf{C}$ with F, K, and E fields and $(E : K) = m$, $(K : F) = n$. Then $(E : F) = mn$. Indeed if $\{u_1, \ldots, u_n\}$ is a basis for K as an F-vector space and if $\{v_1, \ldots, v_m\}$ is a basis for E as a K-vector space, then*

$$\{u_i v_j : 1 \leq i \leq n, 1 \leq j \leq m\}$$

is a basis for E as an F-vector space. In particular $(K : F)$ divides $(E : F)$.

Proof. We need to prove that $\{u_i v_j\}$ is a linearly independent spanning set for E as an F-vector space.

First if $v \in E$, then we may write $v = \sum_{j=1}^{m} a_j v_j$ with $a_j \in K$. Each a_j may in turn be written as an F-linear combination of the u_i's; e.g.,

$$a_j = \sum_{i=1}^{n} b_{ij} u_i.$$

Now substituting we get

$$v = \sum_{i,j} b_{ij} u_i v_j,$$

as desired. Thus $\{u_i v_j\}$ is a spanning set for E as an F-vector space.

Next suppose that $\sum_{i,j} c_{ij} u_i v_j = 0$ for some $c_{ij} \in F$. For each j, set

$$k_j = \sum_{i=1}^{n} c_{ij} u_i.$$

Then

$$\sum_{j=1}^{m} k_j v_j = 0$$

with $k_j \in K$. Since $\{v_1, \ldots, v_m\}$ is a K-linearly independent set, it follows that $k_j = 0$ for all j. But then for each j, we have

$$\sum_{i=i}^{n} c_{ij} u_i = 0$$

with the $c_{ij} \in F$. Since $\{u_1, \ldots, u_n\}$ is an F-linearly independent set, we have $c_{ij} = 0$ for all i, j. Thus $\{u_i v_j\}$ is an F-linearly independent set and hence an F-basis for E, as desired. ∎

This theorem has many important consequences, as we shall see. However, it is important to note that if c is a "random" complex number and F is a subfield of $\mathbf{C}$, then in general $F(c)$ is *not* finite dimensional as a vector space over F. For instance $\mathbf{Q}(\pi)$ is not finite dimensional as a vector space over $\mathbf{Q}$.

Exercises

Throughout let F be a subfield of $\mathbf{C}$.

18.3. Prove that if $(F(c) : F) = n$, then c is a root of an irreducible polynomial of degree n in $F[x]$.

18.4. Let $a, b \in F$. Prove that $E = F(\sqrt{a} + \sqrt{b})$ also contains $\sqrt{a}$ and $\sqrt{b}$. (*Hint*: Compute $(\sqrt{a} + \sqrt{b})^2$ and conclude that $\sqrt{ab} \in E$. Then compute $\sqrt{ab}(\sqrt{a} + \sqrt{b})$.)

18.5. Prove: If $(E : F) = p$ with p a prime, then $E = F(a)$ for each $a \in E - F$.

18.6. Suppose a and b are complex numbers with $(F(a) : F) = m$ and $(F(b) : F) = n$. Suppose further that $\gcd(m, n) = 1$. Prove that $(F(a, b) : F) = mn$ and $F(a) \cap F(b) = F$.

18.7. Prove: If $p(x)$ is an irreducible polynomial of degree n in $F[x]$ and $p(x)$ has a root in the extension field E of F, then n divides $(E : F)$.

18.8. Prove: If $p(x)$ is an irreducible polynomial of degree n in $F[x]$ and if E is an extension field of F with $\gcd(n, (E : F)) = 1$, then $p(x)$ is irreducible in $E[x]$ as well.

18.9. Prove: If $p(x)$ is a polynomial of degree n in $F[x]$ and if E is the splitting field of $p(x)$ over F, then $(E : F)$ is a divisor of $n!$.

18.10. Consider the cyclotomic polynomial $\Phi_7(x) = (x^7 - 1)/(x - 1)$. Let

$$\alpha = \cos\left(\frac{2\pi}{7}\right) + i \sin\left(\frac{2\pi}{7}\right).$$

18.10a. Prove that $\mathbf{Q}(\alpha)$ is the splitting field for $\Phi_7(x)$ over $\mathbf{Q}$.

18.10b. Let $r = \alpha + \bar{\alpha} = \alpha + (1/\alpha)$. Find a cubic polynomial $p(x) \in \mathbf{Q}[x]$ having r as a root. Prove that $p(x)$ is irreducible in $\mathbf{Q}[x]$.

18.10c. Prove that $(\mathbf{Q}(\alpha) : \mathbf{Q}) = 6$.

18.10d. Prove that $\mathbf{Q}(r)$ is the splitting field for $p(x)$ over $\mathbf{Q}$. (*Hint*: Prove that $\mathbf{Q}(r) = \mathbf{Q}(\alpha) \cap \mathbf{R}$.)

18.11. For every divisor m of $6 = 3!$ give an example of a polynomial $p_m(x) \in \mathbf{Q}[x]$ such that the splitting field E_m for $p_m(x)$ over $\mathbf{Q}$ satisfies $(E_m : \mathbf{Q}) = m$. ∎

Armed with Lagrange's Theorem for Fields, we return to the theme of constructible numbers in the sense of straight-edge and compass constructions. Let us note the connection between quadratic field extensions and the constructible numbers.

Definition. Let us call a line an *F*-**line** if it passes through two "points" in F. (Remember we are identifying complex numbers with points in the Euclidean plane.) Let us call a circle an *F*-**circle** if its center is a point in F and its radius is a number in F. Finally let us call a number **one-step constructible** over F if it is the point of intersection of two F-lines or of two F-circles or of an F-line and an F-circle.

Theorem. *If a number r is one-step constructible over F then either $F(r) = F$ or $F(r)$ is a quadratic field extension of F or $F \le F(s) \le F(r)$ with $F(s)$ a quadratic field extension of F and $F(r)$ a quadratic field extension of $F(s)$.*

Proof. If $r = a+bi$ is one-step constructible over F, then (a, b) is a simultaneous solution of a pair of equations, each of the form

$$cx + dy = e$$

or

$$(x - h)^2 + (y - k)^2 = r^2$$

with c, d, e, h, k, r all in F. If both equations are linear, then clearly a and b are in F and so $r \in F$ or $r \in F(i)$, a quadratic extension of F. If both equations are of the second type, then subtracting the second from the first gives a linear equation and so either we may solve for y in terms of x or we may solve for x in terms of y.

Thus we have, for instance, that a is the root of a quadratic equation with coefficients in F and $b \in F(a)$. Thus we have "at worst" $F \leq F(a) \leq F(a)(i) = F(r)$ with each extension quadratic over F. ∎

Remark. *I'm not actually sure if a "one-step" construction can really lead to a "two-step" quadratic extension, but this won't be important in the sequel. The important thing is that a number is constructible if and only if it can be achieved via a finite sequence of one-step constructions and so we get the following characterization of constructible numbers.*

Theorem. *A number r is constructible (over $\mathbf{Q}$) if and only if there is a sequence $\mathbf{Q} \leq \mathbf{Q}(r_1) \leq \mathbf{Q}(r_1, r_2) \leq \cdots \leq \mathbf{Q}(r_1, r_2, \ldots, r_n)$ of field extensions with each field in the sequence a quadratic extension of the preceding field in the sequence and with $r \in \mathbf{Q}(r_1, r_2, \ldots, r_n)$.*

Proof. By the previous theorem and the meaning of constructible number, it is clear that if r is constructible the indicated sequence of fields exists.

Conversely suppose that the indicated sequence of fields exists with r contained in the biggest one. It suffices to argue that if every element of a field F is constructible and if $F \leq F(r)$ with $F(r)$ quadratic over F, then every element of $F(r)$ is constructible. Clearly it is enough to show that r is constructible and we may assume that $r^2 \in F$. But we know that we can construct square roots of constructible numbers. So we are done. ∎

Caution: Suppose that $F \leq F(r) \leq F(r, s)$ with $F(r)$ quadratic over F and with $F(r, s)$ quadratic over $F(r)$. Then $F(r, s)$ is the root field of some quadratic polynomial with coefficients in $F(r)$. But $F(r, s)$ *may not be* the splitting field of *any* polynomial with coefficients in F.

Proof. Let s be a real fourth root of 2. Then $s^2 = \sqrt{2}$ and $\mathbf{Q}(\sqrt{2})$ is quadratic over $\mathbf{Q}$. Also s satisfies the quadratic equation $x^2 - \sqrt{2} = 0$ with coefficients in $\mathbf{Q}(\sqrt{2})$. But $\mathbf{Q}(s)$ is not the splitting field of any polynomial with coefficients in $\mathbf{Q}$, because as we shall see later, if it were, $\mathbf{Q}(s)$ would have to contain all the roots of $x^4 - 2$. But $\mathbf{Q}(s)$ is a real field and the roots of $x^4 - 2$ are $s, -s, is$, and $-is$. ∎

Corollary. *If a number r is constructible over $\mathbf{Q}$, then the minimum polynomial of r over $\mathbf{Q}$ has degree a power of 2.*

Proof. As each extension in the tower from $\mathbf{Q}$ to $\mathbf{Q}(r_1,\ldots,r_n)$ has degree 2, the Lagrange Theorem for field extensions implies that $(\mathbf{Q}(r_1,\ldots,r_n) : \mathbf{Q}) = 2^n$. Since $r \in \mathbf{Q}(r_1,\ldots,r_n)$, 18.7 implies that the degree of the minimum polynomial of r divides 2^n, hence is a power of 2. ∎

As we shall see later, the converse statement is false. There exist numbers r whose minimum polynomial over $\mathbf{Q}$ has degree 4 (for example) but that are not constructible over $\mathbf{Q}$. The precise characterization of constructible numbers depends on a more subtle invariant than the degree of the minimum polynomial: the Galois group of the minimum polynomial. This will be the subject of a later section.

Exercises

18.12. Prove that a 60° angle cannot be trisected by a straight-edge and compass construction. (*Hint*: If this trisection were possible, then the number cos(20°) would be constructible. Why isn't it?)

18.13. Prove that the set K of all constructible numbers is a subfield of $\mathbf{C}$ and that if $a \in K$, then $\sqrt{a} \in K$.

18.14. Prove that the regular heptagon (7-gon) is not constructible.

18.15. Prove that the regular nonagon (9-gon) is not constructible.

18.16. (Classical Greek Problem) Supposedly at the time of some plague, certain Greeks consulted the oracle at Delphi, who ordered them to construct a cubical altar to Apollo twice the volume of the existing cubical altar. This is the reputed source of the problem to "double" a cube using only straight-edge and compass. Prove that this is impossible.

18.17. Find an alternative strategy to avert plagues. ∎

A final famous Greek construction problem was the problem of "squaring the circle" by straight-edge and compass construction, i.e., constructing a square whose area is identical to that of a given circle. This amounts to the problem of constructing $\sqrt{\pi}$. In the 18th century Lambert proved that π is an irrational number. This did not yet settle the constructibility question, since many square roots of irrational numbers are constructible. Finally in the late 19th century Lindemann proved that not only is π not a constructible number, it is in fact a **transcendental number**, i.e., π is not a root of any polynomial equation with coefficients in $\mathbf{Q}$. Thus $\mathbf{Q}(\pi)$ is infinite-dimensional over $\mathbf{Q}$.

CHAPTER 19

GALOIS FIELDS AND THE FUNDAMENTAL THEOREM OF ALGEBRA REVISITED

God created the natural numbers, and all the rest is the work of man.

—Leopold Kronecker

In Chapter 17 we saw how to construct $\mathbf{Q}(\eta)$ where η is the complex root of some polynomial with coefficients in $\mathbf{Q}$ and we saw two ways to prove that $\mathbf{Q}(\eta)$ is a field. In order to complete Euler's proof of the Fundamental Theorem of Algebra, we need more. We need to be able to conjure a root of a polynomial out of thin air. We need to be able to construct the field $\mathbf{Q}(\eta)$ without having any a priori assurance that η exists at all.

In 1831 Galois wrote a note on fields of numbers hinting at a construction that was later clarified and elaborated by Leopold Kronecker. It will enable us to complete Euler's proof of the Fundamental Theorem of Algebra and later it will be very useful in our development of Galois Theory.

In Chapter 14A, we introduced the concept of (2-sided) ideals, left ideals, and quotient rings in the context of noncommutative (or not necessarily commutative) rings. Since the chapter was optional, we redevelop some of these fundamental ideas here in the context of commutative rings.

Definition. Let $(R, +, \cdot)$ be a commutative ring. If I is a subset of R, we say that I is an **ideal** of R if $(I, +)$ is a subgroup of $(R, +)$ and also

$$ra \in I \text{ for all } r \in R \text{ and } a \in I.$$

Exercises

Let R be a commutative ring.

19.1. Verify that $\{0\}$ is an ideal of R.

19.2. Verify that if I and J are two ideals of R, then $I \cap J$ is also an ideal of R.

19.3. Verify that if $1 \in R$, then R is the only ideal of R containing 1.

19.4. Verify that if $a \in R$, then the set $(a) = \{ra : r \in R\}$ is an ideal of R. (*Note*: (a) is called the **principal ideal** of R generated by a.)

19.5. Show by example that in general (a) is larger than the cyclic subgroup $\langle a \rangle$ of $(R, +)$ generated by a. Verify, however, that they are the same if $R = \mathbf{Z}$. ∎

We shall primarily be interested in principal ideals in the sequel.

Definition. Let R be a commutative ring and let I be an ideal in R. We define the **quotient ring** R/I to be the set $\{r + I : r \in R\}$ of all distinct cosets of I with addition and multiplication defined by

$$(a + I) + (b + I) = (a + b) + I$$

and

$$(a + I)(b + I) = ab + I$$

for all a and b in I.

Exercises

19.6. Verify that $(a + I) + (b + I)$ and $(a + b) + I$ are identical as subsets of R.

19.7. Verify that $(a + I)(b + I)$ is a subset of the set $ab + I$. Give an example to show that they are not in general equal. ∎

Exercise 19.7 has the important consequence that the multiplication rule in the definition of R/I is well defined, independent of the choice of element a in the coset $a + I$ and b in the coset $b + I$. Indeed 19.7 shows that multiplication may be defined as follows: Let I_1 and I_2 be two cosets of I. Then $I_1 \cdot I_2$ is the unique coset J of I such that the subset J of R contains the subset $I_1 I_2$ of R.

It is now "obvious" that these new operations satisfy all of the associative laws, commutative laws, and distributive laws that hold in R. And so R/I is a commutative ring. (*However, note*: Even if R is a domain, R/I is in general not a domain.)

We have already seen some of the most important examples of this construction:

The ring $\mathbf{Z}/n\mathbf{Z}$ is the quotient ring of $\mathbf{Z}$ by the principal ideal generated by n.

The ring $\mathbf{Z}[i]/p\mathbf{Z}[i]$ is the quotient ring of $\mathbf{Z}[i]$ by the principal ideal generated by p.

Let's look at one more example that is suggestive for Galois' construction.

Let $\mathbf{R}[x]$ be the domain of all polynomials with real coefficients. Let $(x^2 + 1)$ be the principal ideal generated by the irreducible polynomial $x^2 + 1$. What is the quotient ring $\mathbf{R}[x]/(x^2 + 1)$?

We need to list the cosets of $(x^2 + 1)$. In the example of $\mathbf{Z}/n\mathbf{Z}$, we obtained the cosets of (n) by taking the smallest positive remainders upon division by n:

$$0, 1, \ldots, n - 1.$$

The same idea works here. The possible remainders are 0 and all the polynomials of degree 0 or 1. So the remainders are precisely the cosets

$$a + bx + (x^2 + 1)$$

for any real numbers a and b. Addition is obvious, but multiplication is a little trickier. To save space, let's define a symbol j and write

$$a + bj = a + bx + (x^2 + 1).$$

Now j^2 is the coset of $(x^2 + 1)$ containing the polynomial x^2. Dividing x^2 by $x^2 + 1$ we get

$$x^2 = 1 \cdot (x^2 + 1) + (-1),$$

so the remainder is -1. Thus $x^2 + (x^2 + 1) = -1 + (x^2 + 1)$, and so

$$j^2 = -1.$$

This all begins to look very familiar.

$$\mathbf{R}[x]/(x^2 + 1) = \{a + bj : a, b \in \mathbf{R}, j^2 = -1\}.$$

We've just rediscovered the complex numbers!

Now if we think "abstractly" about what we have just done, we see that we started with a field $\mathbf{R}$ in which the equation

$$x^2 + 1 = 0$$

has no solutions and by doing the quotient ring construction of $\mathbf{R}[x]/(x^2 + 1)$, we have "created" a ring (the complex numbers) in which

$$x^2 + 1 = 0$$

does have a solution (in fact two solutions).

This suggests a way to fix the problem with Euler's proof of the Fundamental Theorem of Algebra. Gauss' objection to Euler's proof was that Euler assumed that there must exist *some* commutative ring of "numbers" containing all the complex numbers in which any given polynomial $p(x)$ with real coefficients has a "full complement" of roots. Maybe this construction will work:

Take the domain $\mathbf{C}[x]$ of polynomials with complex coefficients.
Take the principal ideal $(p(x))$ generated by the polynomial $p(x)$.
Form the quotient ring $\mathbf{C}[x]/(p(x))$.

With luck this will give a field containing $\mathbf{C}$ in which $p(x)$ has a full set of roots.

Actually this is too optimistic for two reasons, but it does get us started in the correct general direction.

To see the first problem, imagine that in our previous example we used the polynomial x^2 instead of $x^2 + 1$. Then again the cosets could all be represented as $a + bk$ for some real numbers a and b and some symbol k such that

$$k = x + (x^2).$$

Exercises

19.8. Describe the multiplication in the ring $\mathbf{R}[x]/(x^2)$. Is this ring a field? What went wrong? What type of element is $x + (x^2)$?

19.9. Describe the multiplication in the ring $\mathbf{R}[x]/(x^2 - x)$. Is this ring a field? What type of element is $x + (x^2 - x)$?

19.10. Prove: If $p(x) = q(x)g(x) \in \mathbf{R}[x]$ with neither q nor g constant polynomials, then $\mathbf{R}[x]/(p(x))$ is not a domain; hence certainly not a field.

19.11. Describe the multiplication in the ring $\mathbf{Q}[x]/(x^2 + x + 1)$. Is this ring a field? What is the multiplicative inverse of $x + (x^2 + x + 1)$? ∎

The good news is: We are OK if $p(x)$ is an irreducible polynomial. Indeed the proof is identical to the proof in Chapter 17 that $\mathbf{Q}(a)$ is a subfield of $\mathbf{C}$.

Theorem. *Let K be a field of numbers and let $p(x)$ be an irreducible polynomial in $K[x]$. Then $K[x]/(p(x))$ is a field containing a copy of K as a subfield.*

Proof. It is always the case that $K[x]/(p(x))$ is a commutative ring with 1. Also the subset

$$K_0 = \{a + (p(x)) : a \in K\}$$

is easily seen to be a subfield of $K[x]/(p(x))$ which is isomorphic to K under the natural mapping $\phi : K \to K_0$:

$$\phi(a) = a + (p(x)).$$

(Since $p(x)$ is not a constant polynomial, the coset $a + (p(x))$ is the 0 coset only if $a = 0$, so the map is one-to-one.)

The only remaining issue is to show that every nonzero element has a multiplicative inverse, so that we can do division in $K[x]/(p(x))$. Again let's set $j = x + (p(x))$. The key fact is that since $p(x)$ is irreducible, if $g(x)$ is any polynomial in $K[x]$, we have the dichotomy that either:

$$g(x) \text{ is a multiple of } p(x), \text{ hence in } (p(x)),$$

or

$$\gcd(p(x), g(x)) = 1.$$

In the first case of the dichotomy, $g(j) = 0$ in $K[x]/(p(x))$, so we don't need to look for an inverse. In the second case, Euclid's Lemma gives polynomials $a(x)$ and $b(x)$ in $K[x]$ with

$$a(x)p(x) + b(x)g(x) = 1$$

and so

$$b(j)g(j) = 1 - a(j)p(j) = 1$$

in $K[x]/(p(x))$. Thus $b(j)$ is the multiplicative inverse of $g(j)$ in the second case, completing the proof. ∎

Exercises

19.12. Find the multiplicative inverse of $j^3 + j$ in the field $\mathbf{Q}[x]/(x^4+x^3+x^2+x+1)$, where $j = x + (x^4 + x^3 + x^2 + x + 1)$.

19.13. What are the solutions of the equation

$$x^4 + x^3 + x^2 + x + 1 = 0$$

in the field $\mathbf{Q}(j) = \mathbf{Q}[x]/(x^4 + x^3 + x^2 + x + 1)$? ∎

Clearly we only have to prove the Fundamental Theorem of Algebra for irreducible polynomials. So we are off and running. Only to hit the second problem! In order to proceed with Euler's proof, we need to have *all* of the roots of our polynomial $p(x)$ in the extended complex number field. But our construction only guarantees the existence of one such root. For example with some pain we can show that in the field $\mathbf{Q}(j) = \mathbf{Q}[x]/(x^3 - 2)$, the polynomial $p(x) = x^3 - 2$ has only *one* root, not three.

We solve this problem by repeating the process. Thus we have the following result.

Theorem. *Let K be a field (of numbers) and let $p(x)$ be a polynomial of degree d in $K[x]$. Then there is a field E containing K such that $p(x)$ has d roots in E.*

Proof. We proceed by induction on d. If $d = 1$, then $p(x) = ax - b$ obviously has one root b/a in K. So we may take $E = K$ and we are done.

Suppose the result is true for all fields K and polynomials of degree less than d. Let $p(x)$ be a polynomial in $K[x]$ of degree d. If p is not irreducible, so that $p(x) = q(x)g(x)$, then by induction, there is a field F containing K in which $q(x)$ has all of its roots and a field E containing F in which $g(x)$ has all of its roots. Then E contains all of the roots of p and we are done.

So assume that p is irreducible. Then $F = K[x]/(p(x))$ is a field containing K in which $p(x)$ has at least one root α. By the Factor Theorem, $p(x) = (x - \alpha)q(x)$ with $q(x) \in F[x]$. Since $q(x)$ has smaller degree, by induction there is a field E containing F and all the roots of $q(x)$. We are done. ∎

Now we can complete Euler's proof of the Fundamental Theorem of Algebra.

Fundamental Theorem of Algebra. *Let $p(x)$ be a polynomial with real coefficients of degree d. Then $p(x)$ has d roots (counting multiplicity) in the complex numbers $\mathbf{C}$.*

Proof. By the Weak Fundamental Theorem, it suffices to show that there is a field of numbers containing $\mathbf{C}$ in which $p(x)$ has d roots. By the preceding theorem, there is a splitting field E for $p(x)$ over $\mathbf{C}$ which does the job. ∎

We now briefly discuss finite fields. These are sometimes known as Galois fields, since the fundamental ideas for their construction were sketched in a paper by Galois. If you haven't done so already, you should read the beginning of Chapter 14A and do the first set of exercises. You will then have the following result.

Theorem. *If F is a finite field, then $|F| = p^n$ for some prime p and $n \in \mathbf{N}$.*

We now wish to prove the converse statement.

Theorem. *If p is a prime and $n \in \mathbf{N}$, then there exists a finite field F with $|F| = p^n$ and F is a splitting field for the polynomial $x^{p^n} - x$.*

Proof. First of all, for every prime p, we have the field $\mathbf{F}_p = \mathbf{Z}/p\mathbf{Z}$. We may consider the polynomial $x^{p^n} - x$ with coefficients in $\mathbf{F}_p$. By the preceding construction, there exists an extension field E of $\mathbf{F}_p$ that contains all the roots of $x^{p^n} - x$. Consider the subset

$$F = \{a \in E : a^{p^n} = a\}.$$

Thus F is the set of all roots of $x^{p^n} - x$, and since $\mathbf{F}_p$ is a field, the Factor Theorem tells us that there are at most p^n distinct roots. Indeed the derivative of $x^{p^n} - x$ is -1, since $p = 0$. Hence by 11.12, $x^{p^n} - x$ has all distinct roots, whence $|F| = p^n$.

We claim that F is a subfield of E, i.e., F is closed under addition, subtraction, multiplication and nonzero division. Let $a, b \in F$. Then

$$(ab)^{p^n} = a^{p^n} b^{p^n} = ab$$

and

$$(a+b)^{p^n} = a^{p^n} + b^{p^n} = a + b, \cdot$$

since all of the binomial coefficients $\binom{p^n}{k}$ are divisible by p for $1 \leq k < p^n$, and hence equal to 0 in E.

Thus F is a splitting field for $x^{p^n} - x$ over $\mathbf{F}_p$ and $|F| = p^n$, as claimed. ∎

Towards the end of Chapter 12, we proved that U_p is a cyclic group for all primes p. The same argument (counting roots of $x^m - 1$ in F to deduce that there must be a primitive root of $x^{p^n-1} - 1$ in F) proves the following theorem.

Theorem. *The multiplicative group $F^\times$ of the finite field F is cyclic, i.e., there exists a primitive p^nth root of 1, λ, in F.*

We now consider the automorphisms of the field F. The following numerical result is useful.

Lemma. *Let p^k be the p-part of the natural number n, i.e., $n = p^k m$ with $gcd(p, m) = 1$. Then $\binom{n}{p^k}$ is not divisible by p.*

Proof. Let $1 \leq r < p^k$ and let $r = p^a m_0$ with $gcd(p, m_0) = 1$. Then $n - r = p^a(p^{k-a}m - m_0)$ and since $k - a > 0$, it follows that p^a is the p-part of $n - r$. Likewise p^a is the p-part of $p^k - r$. Moreover by hypothesis, p^k is the common p-part of n and p^k.

Since
$$\binom{n}{p^k} = \frac{n(n-1)\cdots(n-(p^k-1))}{p^k(p^k-1)\cdots(p^k-(p^k-1))},$$
the lemma is true. ■

Theorem. *Let F be a finite field with $|F| = p^n$. Let A be the group of all field automorphisms of F. Then A is a cyclic group of order n generated by the map $\sigma : F \to F$ defined by $\sigma(x) = x^p$ for all $x \in F$. Moreover for all i, $1 \le i \le n$,*
$$C_F(\sigma^i) = F_i = \{x \in F : x^{p^i} - x = 0\}$$
is the unique subfield of F of cardinality p^i. Moreover $F_1, F_2, \ldots, F_n = F$ are all the subfields of F.

Proof. Since $F^\times = \langle \lambda \rangle$, every group automorphism σ of the group $F^\times$ is given by
$$\sigma(x) = x^i$$
for some $i \in \mathbb{N}$ and all $x \in F^\times$. By taking remainders, we may assume that $1 \le i \le p^n$, since $x^{p^n} = x$ for all $x \in F$. If σ is a field automorphism, then σ must also satisfy
$$\sigma(x+y) = \sigma(x) + \sigma(y)$$
for all $x, y \in F$. In particular,
$$\sigma(x+1) = (x+1)^i = \sigma(x) + \sigma(1) = x^i + 1$$
for all $x \in F$. Subtracting we get
$$p(x) = \binom{i}{1}x^{i-1} + \binom{i}{2}x^{i-2} + \cdots + \binom{i}{i-1}x = 0$$
for all $x \in F$. Since $p(x)$ has degree less than p^n but $p(x)$ has p^n roots in the field F, $p(x) \equiv 0$. Write $i = p^a j$ with $a \ge 0$, $\gcd(p, j) = 1$. Then by the lemma, $\binom{i}{p^a} \ne 0$ in F. It follows that $i = p^a$.

On the other hand it is clear that the map
$$\sigma(x) = x^p \text{ for all } x \in F$$
is a field automorphism of F and so A is a cyclic group generated by σ, as claimed. Furthermore it is clear that for each i, $F_i = C_F(\sigma^i)$ is the set of all roots of $x^{p^i} - x$, which is the unique subfield of F of cardinality p^i. As each subfield of F has cardinality p^j for some j with $1 \le j \le n$, it follows that the F_i are all of the subfields of F. ■

We remark that the map
$$\langle \sigma^i \rangle \to C_F(\langle \sigma^i \rangle)$$
defines a bijection between the subgroups of A and the subfields of F. This is an analogue of the Galois Correspondence which we shall develop in the next two chapters for finite extensions of $\mathbb{Q}$. There is a general context of separable extensions fields which can be developed and which covers both of these examples.

Exercises

We let F be a finite field of order p^n with $F^\times = \langle \lambda \rangle$, and we let A be the automorphism group of F with $A = \langle \sigma \rangle$, where $\sigma(x) = x^p$ for all $x \in F$. We let r be a prime divisor of $p^n - 1$ and let r^a denote the r-part of $p^n - 1$.

19.13. Prove: If $1 \leq k \leq a$, then $C_k = \{x \in F : x^{r^k} = 1\}$ is a cyclic subgroup of $F^\times$ of cardinality r^k, and $\sigma(C_k) = C_k$.

19.14. Suppose that $n = rm$ and $\tau = \sigma^m$.

19.14a. Prove: τ is an automorphism of F of order r and $C_F(\tau) = F_m$ is the subfield of order p^m.

19.14b. Suppose that r^b is the r-part of $p^m - 1$ with $b \geq 1$. Prove that $a \geq b + 1$ with equality if $b \geq 2$. (*Hint*: First "write out" $\frac{p^n - 1}{p^m - 1}$. Using the fact that $p^m \equiv 1 \pmod{r}$, argue that $\frac{p^n - 1}{p^m - 1} \equiv 0 \pmod{r}$, and also that if $p^m \equiv 1 \pmod{r^2}$, then $\frac{p^n - 1}{p^m - 1} \equiv r \pmod{r^2}$.)

19.14c. Suppose that r is odd and r^b is the r-part of $p^m - 1$ with $b \geq 1$. Prove that $a = b + 1$. (*Hint*: By part b, you may assume that $b = 1$ and $a \geq 3$. Let C_3 be as in 19.13. Argue that $\tau(C_3) = C_3$ and $|C_{C_3}(\tau)| = r$. Argue that this contradicts 16.29d, since r is odd.)

19.14d. Give an example to show that 19.14c would be false if $r = 2, b = 1$. ∎

APPENDIX: SYLOW'S THEOREMS, ZSIGMONDY'S THEOREM AND WEDDERBURN'S THEOREM

The results we have developed above provide all the necessary tools to prove several important and celebrated theorems. We begin with the fundamental theorems of Ludwig Sylow (1873).

Sylow's Theorems. *Let G be a finite group and p a prime. Let p^a be the p-part of $|G|$ and let $\mathcal{P}$ denote the set of all subgroups of G of cardinality p^a. Then the following conclusions hold:*

1. $\mathcal{P}$ *is nonempty, i.e., G has at least one subgroup of cardinality p^a (in honor of Sylow, subgroups of G of cardinality p^a are called Sylow p-subgroups of G);*
2. G *permutes the set $\mathcal{P}$ transitively via conjugation;*
3. $|\mathcal{P}| \equiv 1 \pmod{p}$*; and*
4. *If R is a subgroup of G with $|R|$ a power of p, then R is contained in some Sylow p-subgroup of G.*

We shall develop the proof in the following exercises.

19.15. Let G be a finite group, p and p^a the p-part of $|G|$. Let $\mathcal{S}$ be the set of all subsets of G of cardinality p^a.

19.15a. Using the preceding lemma on binomial coefficients, prove that $|\mathcal{S}|$ is not divisible by p.

19.15b. Prove: If $g \in G$ and $X \in \mathcal{S}$, then $gX \in \mathcal{S}$ and the map $\pi_g : \mathcal{S} \to \mathcal{S}$ via $\pi_g(X) = gX$ defines a permutation of the set $\mathcal{S}$ for all $g \in G$. Moreover if $\Pi : G \to Sym(\mathcal{S})$ is the map

$$\Pi(g) = \pi_g,$$

then Π is a homomorphism of groups. Thus G acts as a group of permutations of the set $\mathcal{S}$.

19.15c. Prove Sylow's Theorem: G has a subgroup of cardinality p^a. (*Hint*: Suppose not. Using the Well-Ordering Principle, choose G to be a counterexample of minimum order. For $X \in \mathcal{S}$, let G_X be the subgroup of G fixing the "point" X. Using 19.15a, argue that we may choose X to lie in a G-orbit $\mathcal{O}$ of cardinality not divisible by p. Using Lagrange's Theorem, conclude that G_X has cardinality divisible by p^a. If $G_X \neq G$, argue that you are done by minimal choice of G. If $G_X = G$, argue that $XX = X$ and conclude that X is a subgroup of cardinality p^a.)

19.16. (Bonus: The rest of Sylow's Theorems) Let $\mathcal{P}$ denote the set of all Sylow p-subgroups of the finite group G.

19.16a. Prove: G acts as a group of permutations of $\mathcal{P}$ via the conjugation action, namely, for $g \in G$, define $c_g : \mathcal{P} \to \mathcal{P}$ via

$$c_g(P) = gPg^{-1} \text{ for all } P \in \mathcal{P}.$$

19.16b. For $P \in \mathcal{P}$, let G_P denote the stabilizer of the "point" P. Prove: If R is a p-subgroup of G and $R \leq G_P$, then $R \leq P$.

19.16c. Using 19.16b and the idea of 13.5, prove that every G-orbit $\mathcal{O}$ on $\mathcal{P}$ has cardinality $|\mathcal{O}| \equiv 1 \pmod{p}$. (*Hint*: Let $P_1 \in \mathcal{O}$ and consider $|Fix(P_1)|$, the number of fixed points of the group P_1 acting on the set $\mathcal{O}$.)

19.16d. Prove: There is only one G-orbit on $\mathcal{P}$ and $|\mathcal{P}| \equiv 1 \pmod{p}$. (*Hint*: Suppose there are at least two orbits, one of which is $\mathcal{O}$, and now choose a Sylow p-subgroup P with $P \notin \mathcal{O}$ and consider the number of fixed points of the group P acting on the set $\mathcal{O}$.) ∎

The following exercise is completely unrelated to everything that goes before, but, it will be used in conjunction with 19.14 to prove Zsigmondy's theorem for odd primes p.

19.17. Let $n \in \mathbf{N}$ and let $\Phi_n(x)$ denote the nth cyclotomic polynomial. Let p and r be primes with $p \neq 2$ and with r a divisor of n. Prove: $|\Phi_n(p)| > r$. (*Hint*: Let $\zeta = e^{2\pi i/n}$. Factor $\Phi_n(x)$ as a product of factors of the form $x - \zeta^i$. Argue that $|p - \zeta^i| > 2$. Then prove that $2^x > x + 1$ for all $x > 1$.) ∎

We now have all the tools necessary to give a short proof of Zsigmondy's Theorem for odd primes p.

Zsigmondy's Theorem. *Let p be an odd prime and let $n \in \mathbf{N}$ with $n > 2$. There exists a prime r such that r divides $p^n - 1$ but r does not divide $p^m - 1$ for all $m \in \mathbf{N}$, $m < n$.*

We proceed in a short sequence of lemmas. We suppose that the theorem is false for some prime p.

Lemma 1. *Let r be a prime divisor of $p^n - 1$ and let m be the smallest natural number such that r divides $p^m - 1$. Then m is a proper divisor of n.*

Proof. m is the order of p in U_r and so m divides n. ∎

Lemma 2. *$n \neq 2^k$ for any k.*

Proof. Since $2^k > 2$, $p^{2^{k-1}} + 1 \equiv 2 \pmod{4}$. Hence there is an odd prime divisor r of $p^{2^{k-1}} + 1$. But then p has order 2^k in U_r, contrary to assumption. ∎

Lemma 3. *Suppose that r is a prime and r^a is the p-part of $p^n - 1$ for some $a > 1$. Suppose that r^a does not divide $p^m - 1$ for any $m < n$. Then the following conclusions hold:*
(a) $n = rk$ for some $k \in \mathbf{N}$;
(b) r is odd;
(c) r^{a-1} divides $p^k - 1$; and
(d) r is unique.

Proof. Since Zsigmondy's Theorem is false by assumption, there exists a proper divisor m of n such that r divides $p^m - 1$. Let r^b be the r-part of $p^m - 1$. Since $b < a$ by hypothesis, we see as in 19.14 that n/m is divisible by r, proving (a). Applying this with m the 2-part of n, we conclude that the 2-part of $p^m - 1$ is the same as the 2-part of $p^n - 1$. Since $m < n$ by Lemma 2, we conclude that r is odd, proving (b). Write $n = rk$. Then by 19.14, r^{a-1} is the r-part of $p^k - 1$. Thus (c) holds.

It remains to prove that r is unique. Suppose on the contrary that r and s both satisfy the hypotheses. Then $n = rsd$ for some $d \in \mathbf{N}$. If $p^d \equiv 1 \pmod{r}$, then arguing as before, $p^{rd} - 1$ has the same r-part as $p^n - 1$, contrary to assumption. Hence $p^d \not\equiv 1 \pmod{r}$, but $p^{sd} \equiv 1 \pmod{r}$. Hence p^d has order s in U_r, whence s divides $r - 1$ by Fermat's Little Theorem. But symmetrically, p^d has order r in U_s, whence r divides $s - 1$, which is impossible. This proves that r is unique. ∎

Corollary 4. *$|\Phi_n(p)| = 1$ or r with r as in Lemma 3.*

Proof. Let D be the set of proper divisors of n and let M be the least common multiple of the members of the set $\{p^m - 1 : m \in D\}$. Then M divides $N = \prod_{m \in D} \Phi_m(p)$. Note that $\Phi_n(p) = (p^n - 1)/N$ and so $\Phi_n(p)$ is a divisor of $(p^n - 1)/M$. By hypothesis and Lemma 3, $(p^n - 1)/M = 1$ or r for some unique odd prime r. Thus the same is true for $|\Phi_n(p)|$.

Thus in particular $|\Phi_n(p)| \leq r$. But this contradicts 19.17, completing the proof of Zsigmondy's Theorem for odd primes p. ∎

As noted in Chapter 14A, Zsigmondy's Theorem for odd p in conjunction with Cauchy's Theorem gives the following case of Wedderburn's Theorem.

Theorem (Wedderburn). *Let D be a finite skew field with $|D|$ odd. Then D is a field.*

CHAPTER 20

GALOIS' THEORY OF EQUATIONS

J'ai fait en Analyse plusieurs choses nouvelles.
I have done several new things in Analysis.

—Évariste Galois (1832)

Let's try to imagine the thought processes of the young genius Évariste Galois as he contemplated the work of his predecessors in the theory of equations.

On the one hand there was the great paper of 1770–1771 by Lagrange in which Lagrange examined the work of *his* predecessors—in particular the successful solutions of the cubic and quartic equations by Cardano, Viete, and Ferrari—and attempted to extract a universal guiding principle. The principle Lagrange discovered was that of "symmetries" of the roots of the polynomial $p(x)$. He let the symmetric group S_n act on the roots and found auxiliary equations (Lagrange resolvents) whose solution would lead to a solution of $p(x)$ itself. However, Lagrange's paper was finally pessimistic. He concluded (although he did not prove) that these methods would never give a recipe for solving equations of degree higher than 4.

On the other hand there was the work of Gauss (published in his *Disquisitiones Arithmeticae* in 1801), in which he showed how to solve the polynomial equation $x^{17} - 1 = 0$ by successive extraction of square roots and indeed gave a general algorithmic procedure for the solution of any equation of the form $x^n - 1 = 0$ by successive extraction of roots. Here the key role (implicitly) was played by Gauss' group G_n, a much smaller group than S_n.

Galois realized that Gauss was on the right track. When considering a specific polynomial $p(x)$, one should not treat its roots as "indeterminates"—$r_1, r_2, \ldots, r_n$—and indiscriminantly apply every possible permutation in S_n to them. One should *remember* the algebraic relationships among the roots and apply only those permutations that respect those relationships. This leads to the following definition.

Definition. Let F be a subfield of $\mathbf{C}$ and let E be the splitting field over F of a polynomial $p(x) \in F[x]$. The **Galois group** of $p(x)$ over F (also called the Galois group of E/F: $Gal(E/F)$) is the set of all functions $\phi : E \to E$ satisfying the following for all $u, v \in E$:

$$\phi(u+v) = \phi(u) + \phi(v),$$
$$\phi(uv) = \phi(u)\phi(v),$$

and

$$\phi(u) = u \text{ whenever } u \in F.$$

Any function $\phi : E \to E$ satisfying these three conditions is called an F-**automorphism** of the field E. More generally if E, and E' are two extension fields of F with $(E : F) = (E' : F)$ and if $\phi : E \to E'$ is a function satisfying the three conditions, then ϕ is called an F-**isomorphism** of the fields E and E'. There are a few elementary but fundamental facts to be established.

Exercises

20.1. Prove: If $\phi : E \to E'$ is an F-isomorphism, then ϕ is a one-to-one function from E onto E'.

20.2. Prove: If $\phi : E \to E'$ is an F-isomorphism of fields, then ϕ is an isomorphism of F-vector spaces.

20.3. Prove: If $\phi \in Gal(E/F)$, then also $\phi^{-1} \in Gal(E/F)$.

20.4. Prove: $Gal(E/F)$ is a group.

20.5. Prove: If $F = \mathbf{Q}$, then the final condition may be replaced by: $\phi(1) \neq 0$.

20.6. Let $\Phi_n(x)$ be the nth cyclotomic polynomial over $\mathbf{Q}$. Prove that the Gauss group G_n is the Galois group of $\Phi_n(x)$ over $\mathbf{Q}$. (*Hint*: Use the irreducibility of $\Phi_n(x)$ in $\mathbf{Q}[x]$. Let $\eta = e^{2\pi i/n}$. Prove that any Galois automorphism g of $\mathbf{Q}(\eta)$ is determined by $g(\eta)$.)

20.7. Let E be the splitting field of some polynomial over F. Suppose there is a Galois automorphism $\phi : E \to E$ and two elements a and b of E with $\phi(a) = b$. Prove that a and b have the same minimum polynomial over F. (Hint: Let $p(x)$ be the minimum polynomial of a and apply ϕ to the equations $p(a) = 0$.) ∎

Other than the identity function $I : E \to E$ defined by $I(x) = x$ for all $x \in E$, it is not clear in general that there are any Galois automorphisms of E. The remarkable fact is that there are quite a few. The main result we shall prove is the following converse of 20.7.

Definition. Let F be a subfield of $\mathbf{C}$. We shall call a field E a *Galois field* over F if E is the smallest (splitting) subfield of $\mathbf{C}$ containing F and all of the roots of some polynomial $p(x) \in F[x]$.

Theorem. *Let E be a Galois field over F. Let $a \in E$ with minimum polynomial $p(x) \in F[x]$. Let b be any root of $p(x)$. Then $b \in E$ and there is a Galois automorphism $\phi : E \to E$ with $\phi(a) = b$.*

We shall prove a short sequence of results leading up to this theorem.

Theorem 1. *Let F and F' be subfields of $\mathbf{C}$ and let $h : F \to F'$ be an isomorphism of fields (i.e., h is a surjective map with $h(a+b) = h(a) + h(b)$ and $h(ab) = h(a)h(b)$ for all $a, b \in F$ and $h(1) \neq 0$). Let a be a root of the irreducible polynomial $p(x) \in F[x]$ and let b be a root of the irreducible polynomial $h(p(x)) \in F'[x]$ (i.e., apply h to each of the coefficients of $p(x)$). Then there is an isomorphism $h^* : F(a) \to F'(b)$ such that $h^*(c) = h(c)$ for all $c \in F$, and $h^*(a) = b$. (We say that h^* extends h to an isomorphism $F(a) \to F'(b)$.)*

Sketch of a Proof. The idea of the proof is easy, based on the Galois-Kronecker construction of field extensions. (We leave the details to the interested reader to work out for her/himself.) By Galois-Kronecker:

$$F(a) \cong F[x]/(p(x)) \text{ and } F'(b) \cong F'[x]/(h(p(x))).$$

Now the isomorphism $h : F \to F'$ extends to an isomorphism $h : F[x] \to F'[x]$ by applying h to each coefficient of a polynomial in $F[x]$. Under this isomorphism, the principal ideal $(p(x))$ is sent to the principal ideal $(h(p(x)))$ and so there is an induced isomorphism:

$$\overline{h} : F[x]/(p(x)) \to F'[x]/(h(p(x))).$$

Now the composition of the three isomorphisms gives an isomorphism:

$$h^* : F(a) \to F(b)$$

via

$$h^*(c_0 + c_1 a + \cdots + c_m a^m) = h(c_0) + h(c_1)b + \cdots + h(c_m)b^m,$$

as desired. ∎

Theorem 2. *Let E be the Galois field over F of the polynomial $p(x) \in F[x]$ and let $\phi : E \to E'$ be an F-isomorphism of subfields of $\mathbf{C}$ containing F. Then ϕ is a permutation of the roots of $p(x)$ and $E' = E$.*

This is an unbelievable theorem. (I still don't believe it.) Taking $F = \mathbf{Q}$, this says that if E is the splitting field of a polynomial $p(x) \in \mathbf{Q}[x]$, then no other subfield of $\mathbf{C}$ is isomorphic (as a field) to E.

Proof. Let α be a root of $p(x)$. Since ϕ is the identity map on F, applying ϕ to the equation

$$p(\alpha) = 0$$

yields the equation

$$p(\phi(\alpha)) = 0.$$

Thus $\phi(\alpha)$ is also a root of $p(x)$ and since ϕ is a one-to-one function on E, it is in particular a permutation of the roots of $p(x)$.

But now by the constructive method of building splitting fields, every number $e \in E$ is expressible as a polynomial in the roots of $p(x)$ with coefficients in F:

$$e = g(\alpha_1, \alpha_2, \ldots, \alpha_n).$$

Since ϕ permutes the α_i's and fixes the coefficients of g, $\phi(e)$ is a "value" of g (in the language of Lagrange) lying in the field E.

Thus $\phi(E) \subseteq E$. But $\phi(E) = E'$. So $E' \subseteq E$. But $(E : F) = (E' : F)$, since ϕ is an isomorphism of F-vector spaces. So $E' = E$, as claimed. ∎

Now we are ready to prove our main theorem.

Main Theorem on Galois Automorphisms. *Let E be a Galois field over F. Let L be any subfield of E, and let $\phi : L \to L'$ be an F-isomorphism of L with some extension field L' of F. Then there is a Galois automorphism $\sigma \in Gal(E/F)$ that extends ϕ; i.e.,*

$$\sigma(x) = \phi(x) \text{ for all } x \in L.$$

In particular L' is a subfield of E.

Thus for any $\alpha \in E$, if $p(x)$ is the minimum polynomial of α over F, then E contains a splitting field for $p(x)$. Furthermore if β is any root of $p(x)$, then there is a Galois automorphism $\sigma \in Gal(E/F)$ with $\sigma(\alpha) = \beta$.

Proof. First notice that if α and β are roots of the same minimum polynomial over F, then by Theorem 1 there is an F-isomorphism $\phi : F(\alpha) \to F(\beta)$ with $\phi(\alpha) = \beta$. Thus if $\alpha \in E$, we see that the last paragraph of the Main Theorem follows immediately from the first paragraph.

Now we prove the first paragraph by complete mathematical induction on $n = (E : L)$. Notice that the case $n = 1$ is precisely Theorem 2. So assume the theorem is true for all subfields K of E with $(E : K) < n = (E : L)$. Since $n > 1$, we may choose $a \in E - L$. Let $f(x)$ be the minimum polynomial for a over L and let b be any root of the polynomial $\phi(f(x)) \in L'[x]$. Then by Theorem 1, ϕ extends to an F-isomorphism $\phi' : L(a) \to L'(b)$. Since $(E : L(a)) < (E : L)$, our inductive hypothesis implies that ϕ' extends to a Galois automorphism $\sigma \in Gal(E/F)$. Obviously σ extends ϕ and we are done. ∎

Exercises

20.8. Let E denote the splitting field of $p(x) = (x^2 - 2)(x^3 - 1)$ over $\mathbf{Q}$. Prove that $(E : \mathbf{Q}) = 4$. Give an argument that $x^2 - 2$ remains irreducible over F, the splitting field of $x^3 - 1$. Prove that $Gal(E/\mathbf{Q})$ is a noncyclic group of order 4. Give three different subfields of E of degree 2 over $\mathbf{Q}$.

20.9. Let E be as in 20.8. Prove that E is also the splitting field of $f(x) = (x^2 - 2x - 2)(x^2 + 1)$.

20.10. Find the splitting field and Galois group for $q(x) = x^3 - 5$ over $\mathbf{Q}$.

20.11. Find the splitting field and Galois group for $g(x) = x^4 - 2x^2 + 9$ over $\mathbf{Q}$.

20.12. Let L be the splitting field of $g(x) = x^4 - 2$. Prove that $(L : \mathbf{Q}) = 8$. Prove that $Gal(L/\mathbf{Q})$ is a subgroup of S_4 of cardinality 8. (*Hint*: Show it has cardinality at most 8. Next show it has cardinality at least 8. Lagrange's Theorem can help.) Conclude that $Gal(L/\mathbf{Q}) \cong D_8$. ∎

CHAPTER 21

THE GALOIS CORRESPONDENCE

Tu prieras publiquement Jacobi ou Gauss de donner leur avis, non sur la verité, mais sur l'importance des théorèmes.
Apres cela, il y aura, j'espere des gens qui trouveront leur profit a déchiffrer tout ce gachis.

[Please ask Jacobi or Gauss to give their opinion publicly, not about the truth, but about the importance of these theorems. After that, I hope there will be some people who will take profit from deciphering all this mess.]

—Évariste Galois

We are finally up to the amazing Correspondence Theorem discovered by Galois relating the internal structure of the Galois field E/F to the internal structure of its Galois group $Gal(E/F)$. One amazing aspect of this theorem is that it describes the internal structure of an *infinite*, albeit finite dimensional, object E in terms of the internal structure of its *finite* group of Galois automorphisms.

Thus in particular we shall see that although E has *infinitely many* subspaces as a vector space over F, E has only *finitely many* subfields containing the field F. This is in fact an easy consequence of our previous results. First note that if E is the splitting field of a polynomial $p(x) \in F[x]$ and if K is any intermediate field between F and E, then E is also the splitting field of $p(x)$ regarded as a polynomial in $K[x]$. Thus it makes sense to speak of the Galois group $Gal(E/K)$.

Exercise

21.1. Prove:
$$Gal(E/K) = \{\sigma \in Gal(E/F) : \sigma(x) = x \text{ for all } x \in K\}. \quad \blacksquare$$

This easy remark has the following important consequence.

Theorem. *Let E be a Galois field over F and let K be any intermediate field. For any $\alpha \in E - K$, there is a Galois automorphism $\sigma \in Gal(E/K)$ with $\sigma(\alpha) \neq \alpha$.*

Proof. Since $\alpha \notin K$, the minimum polynomial $f(x)$ of α over K has degree at least 2. By an earlier exercise, $f(x)$ does not have a multiple root (since otherwise f and its derivative f' would have a common factor in $K[x]$, contrary to the irreducibility of f) and so $f(x)$ has a root $\beta \neq \alpha$. Then by the Main Theorem on Galois Automorphisms, there exists $\sigma \in Gal(E/K)$ with $\sigma(\alpha) = \beta$, as claimed. $\blacksquare$

Now we set up the fundamental Galois Correspondence:

Let K be a subfield of E containing F.

Let H be a subgroup of $Gal(E/F)$.

Set
$$E^H = \{e \in E : \sigma(e) = e \ \forall \ \sigma \in H\}.$$

The Galois Correspondence associates to the field K the group $Gal(E/K)$ and to the group H the field E^H:

$$K \implies Gal(E/K)$$
$$H \implies E^H$$

Our goal is to show that these two "arrows" are inverses of each other. We have already established this in one direction.

Exercises

21.2. Prove: Let K be any subfield of E containing F. Then
$$K = E^{Gal(E/K)}.$$

21.3. Prove: There are at most as many subfields of E containing F as there are subgroups of $Gal(E/F)$. In particular there are only finitely many subfields of E containing F. (If $(E : F) = n$, what is an upper bound on $|Gal(E/F)|$? What is an upper bound on the number of subgroups of $Gal(E/F)$?) $\blacksquare$

In order to determine exactly how big $Gal(E/F)$ is and to complete the proof of the Galois Correspondence Theorem, we shall establish the fundamental Primitive Element Theorem of Galois. We shall state and prove the following version, which is all that we need. There exist slightly stronger versions.

Primitive Element Theorem of Galois. *Let E/F be a Galois extension with E a subfield of $\mathbf{C}$. There exists $\alpha \in E$ (called a **primitive element**) such that $E = F(\alpha)$.*

There are different ways to prove the Primitive Element Theorem. We shall establish it via a short and elementary digression into linear algebra. Throughout we shall assume that F is an infinite field (e.g., a subfield of the complex numbers) and V is a vector space over F of dimension n.

Definition. A subspace H of V is called a **hyperplane** of V if H is $(n-1)$-dimensional over F.

Definition. A linear transformation $\phi : V \to F$ is called a **linear functional** on V. The set of all linear functionals on V is called the **dual space** V^* of V.

Exercises

21.4. Prove: V^* is an F-vector space under pointwise addition of linear functionals.

21.5. Prove: V^* is n-dimensional as an F-vector space. Indeed if $\{e_1, \ldots, e_n\}$ is an F-basis for V, then V^* has the **dual basis** $\{e_1^*, \ldots, e_n^*\}$, defined by

$$e_i^*(e_j) = 0 \text{ if } j \neq 1 \text{ and } e_i^*(e_i) = 1.$$

21.6. Prove: The kernel (null-space) of a linear functional is a hyperplane of V. Moreover the correspondence

$$H \iff \{f \in V^* : Ker(f) = H\}$$

is a one-to-one correspondence between the set of all hyperplanes of V and the set of all one-dimensional subspaces of V^*.

21.7. Prove: If F is an infinite field, then V^* has infinitely many one-dimensional subspaces, and so V has infinitely many hyperplanes. (*Hint*: Prove that in the plane spanned by e_1^* and e_2^*, there are infinitely many vectors, no two of which are collinear.) ■

Theorem. *Let V be a vector space of dimension $n \geq 1$ over the infinite field F. Then V is not the union of a finite collection of proper subspaces.*

Proof. We proceed by induction on n. If $n = 1$, then $\{0\}$ is the only proper subspace of V and so the result is obvious.

Now assume that the result is true for all vector spaces of dimension $n - 1$ over F, with $n \geq 2$. Suppose the result is false for V. Since every proper subspace of V is contained in a hyperplane of V, we have that

$$V = H_1 \cup H_2 \cup \cdots \cup H_r$$

for some set $\mathcal{H} = \{H_1, H_2, \ldots, H_r\}$ of hyperplanes of V.

By 21.7 V has infinitely many hyperplanes and so we may choose a hyperplane H of V not in this finite set. Then

$$H = (H \cap H_1) \cup (H \cap H_2) \cup \cdots \cup (H \cap H_r).$$

If $H = H \cap H_i$ then $H = H_i$, since both H and H_i have the same dimension. But this is contrary to the choice of H. So $H \cap H_i$ is a proper subspace of H for all i. But then H is the union of a finite collection of proper subspaces of H, contrary to the inductive assumption. This completes the proof. ∎

Now we can easily prove the Primitive Element Theorem. Let's repeat the statement.

Primitive Element Theorem of Galois. *Let E/F be a Galois extension with E a subfield of $\mathbf{C}$. There exists an element $\alpha \in E$ (called a **primitive element**) such that $E = F(\alpha)$.*

Proof. Suppose that α is not a primitive element of E. Then $F(\alpha)$ is a proper subfield of E. If this is true for every $\alpha \in E$, then E is the union of its proper subfields. But we know by 21.3 that E has only finitely many proper subfields and each one is a proper subspace of the nonzero F-vector space E, contradicting our linear algebra theorem. Thus E contains (many) primitive elements. ∎

Now we can quickly finish the fundamental Galois Correspondence Theorem.

Corollary 1. $|Gal(E/F)| = (E : F)$.

Proof. Let α be a primitive element of E/F, i.e., $E = F(\alpha)$. Then the minimum polynomial $p(x)$ of α over F has degree $n = (E : F)$. Let σ be any Galois automorphism of E/F. Since every element of E is an F-polynomial in α, σ is completely determined by $\sigma(\alpha)$. Now $\sigma(\alpha)$ is one of the n roots of $p(x)$ and so

$$|Gal(E/F)| \leq n = (E : F).$$

On the other hand, by the Main Theorem on Galois Automorphisms, for every root β of $p(x)$, there exists one (and hence only one) Galois automorphism $\sigma \in Gal(E/F)$ with $\sigma(\alpha) = \beta$. Thus

$$|Gal(E/F)| = n = (E : F),$$

as claimed. ∎

Corollary 2. *Let H be any subgroup of $Gal(E/F)$. Then*

$$H = Gal(E/E^H).$$

Proof. Let

$$H^* = Gal(E/E^H) = \{\sigma \in Gal(E/F) : \sigma(x) = x \text{ for all } x \in E^H\}.$$

By the definition of E^H, we know that $H \subseteq H^*$. Also by Corollary 1, we know that

$$|H^*| = (E : E^H).$$

We must show that $(E : E^H) \leq |H|$.

Let $H = \{1 = h_1, h_2, \ldots, h_m\}$, and let α be a primitive element of E/F. Set

$$g(x) = (x - \alpha)(x - h_2(\alpha)) \cdots (x - h_m(\alpha)).$$

The coefficients of $g(x)$ are the elementary symmetric functions in

$$\{\alpha = h_1(\alpha), h_2(\alpha), \ldots, h_m(\alpha)\}$$

and so they are fixed by every function in H. Thus they are in E^H by the definition of E^H.

Since α is a root of $g(x) \in E^H[x]$, the minimum polynomial $f(x)$ of α over E^H has degree at most $m = deg(g)$. Since $E = E^H(\alpha)$, $(E : E^H) = deg(f) \leq m = |H|$, as claimed. ∎

Thus we have established the Fundamental Theorem of Galois Theory.

Fundamental Theorem of Galois Theory. *Let F be a subfield of $\mathbf{C}$. Let E be a Galois field (splitting field) over F. Then the correspondence*

$$K = E^H \iff H = Gal(E/K)$$

defines a one-to-one inclusion-reversing correspondence between the (lattice of) subfields of E containing F and the (lattice of) subgroups of $Gal(E/F)$. Moreover $|H| = (E : E^H)$ for every subgroup H of $Gal(E/F)$.

Exercises

21.8. For each of the equations listed, determine the Galois group over $\mathbf{Q}$ of the equation. List all of the subgroups of the Galois group. List all of the subfields of the splitting field of the equation, and draw a diagram illustrating the Galois correspondence between subgroups and subfields:
 (a) $(x^2 + 1)(x^2 - 2)$
 (b) $(x^2 - 2)(x^2 - 3)(x^2 + 1)$ (*Note*: You must prove by explicit calculation that $\sqrt{3}$ is not in $\mathbf{Q}(\sqrt{2})$.)
 (c) $x^3 - 2$
 (d) $x^7 - 1$
 (e) $x^4 - 3$
 (f) $x^{11} - 1$

21.9. For each finite group G of cardinality at most 7, give an example of an equation whose Galois group over $\mathbf{Q}$ is isomorphic to G. ∎

CHAPTER 22

CONSTRUCTIBLE NUMBERS AND SOLVABLE EQUATIONS

Nils Henrik Abel realized that the "method of Mr. Gauss" (as Galois called it) worked because the Galois groups of the cyclotomic polynomials satisfy the Commutative Law for Composition of Permutations. Following up on this insight, Abel extended the applications of the method of Mr. Gauss to give a formula for dividing the lemniscate into n equal arcs (analogous to inscribing the regular n-gon in the circle). In honor of his insight we call groups whose operation satisfies the commutative law **abelian groups**.

Galois achieved a deeper insight, recognizing that the method of Mr. Gauss depends on a certain permutability condition for subsets of the Galois group, which is analogous to the Commutative Law for Elements.

Definition. A subgroup H of a group G is called a **normal subgroup** of G if
$$gH = Hg \text{ for all } g \in G.$$

Normal subgroups are intimately connected to splitting fields. We continue to assume as usual that F is a subfield of the complex numbers and E is a Galois field extension of F. We let $G = Gal(E/F)$. For any subfield K of E containing F and any $\sigma \in G$, we let
$$\sigma(K) = \{\sigma(k) : k \in K\}.$$

Theorem. *Let K be a subfield of E containing F. Then K has a primitive element β. Moreover K is a splitting field over F if and only if $\sigma(K) = K$ for all $\sigma \in G$.*

Proof. As K has only finitely many subfields containing F, the usual proof shows that K has a primitive element β. Now for all $\sigma \in G$, $\sigma(\beta)$ is a root of the minimum polynomial $p(x)$ of β over F. Thus if K is a splitting field, then $\sigma(\beta) \in K$ and so $\sigma(K) = K$. On the other hand, if $\sigma(K) = K$ for all $\sigma \in G$ and if γ is any root of $p(x)$, then $\gamma = \tau(\beta)$ for some $\tau \in G$ and so $\gamma \in K$, whence K is the splitting field for $p(x)$ over F. ∎

Exercises

22.1. Prove: Let G be a group and H a subgroup of G. Then H is a normal subgroup of G if and only if
$$\sigma H \sigma^{-1} = H \text{ for all } \sigma \in G.$$

22.2. Prove: Let G be a group and H a normal subgroup of G. Then the set $\{gH : g \in G\}$ of all left cosets of H in G can be given the structure of a group (called the **quotient group** G/H) by defining the multiplication
$$(aH)(bH) = abH.$$

22.3. Prove: Let $G = Gal(E/F)$. For any subgroup H of G and any element $\sigma \in G$, we have
$$\sigma(E^H) = E^{\sigma H \sigma^{-1}}.$$

22.4. Prove: If G is a finite group in which every element g satisfies $g^2 = I$, then G is an abelian group of cardinality 2^n for some $n \geq 0$. (*Hint*: By 13.14a, G is an abelian group. Consider a minimal set of generators for G: $a, b, c, \ldots$. Argue that every element of G is of the form $a^i b^j c^k \ldots$ with $i, j, k, \ldots$, all in $\{0, 1\}$.) ∎

Now we can refine the Galois Correspondence as follows.

Theorem. *Let H be any subgroup of $G = Gal(E/F)$. Then E^H is a Galois field over F if and only if H is a normal subgroup of G. (In this case $Gal(E^H/F) \cong G/H$.)*

Proof. By the previous theorem, E^H is a Galois field over F if and only if $\sigma(E^H) = E^H$ for all $\sigma \in G$. By 22.3,
$$\sigma(E^H) = E^{\sigma H \sigma^{-1}}.$$

So by the Galois Correspondence Theorem, E^H is a Galois field over F if and only if $H = \sigma H \sigma^{-1}$ for all $\sigma \in G$ and this holds if and only if H is a normal subgroup of G (using 22.1). (It is not hard to show that $Gal(E^H/F) \cong G/H$. We leave this to the motivated reader to work out for him/herself.) ∎

Now we can prove the following characterization of constructible numbers.

Theorem. *The complex number α is constructible over $\mathbf{Q}$ if and only if the Galois group G of the minimum polynomial of α over $\mathbf{Q}$ is a 2-group; i.e., $|G| = 2^n$ for some $n \geq 0$.*

We prove this result in two parts.

Part 1. *If α is constructible over $\mathbf{Q}$, then the Galois group G of the minimum polynomial of α over $\mathbf{Q}$ is a 2-group.*

Proof. Since α is constructible over $\mathbf{Q}$, there is a tower of subfields of $\mathbf{C}$:

$$\mathbf{Q} = F_0 \subset F_1 \subset \cdots \subset F_n$$

such that $\alpha \in F_n$ and $(F_{i+1} : F_i) = 2$ for all i. For each F_i, let K_i be the smallest Galois subfield of $\mathbf{C}$ containing F_i. (It is clear that K_i exists and $(K_i : \mathbf{Q}) < \infty$.) Then we get a new tower:

$$\mathbf{Q} = K_0 \subset K_1 = F_1 \subset K_2 \subseteq \cdots \subseteq K_n.$$

We claim that $(K_n : \mathbf{Q})$ is a power of 2. If this is the case, then we are done, since K_n contains the splitting field K for α over $\mathbf{Q}$ and so by the Lagrange Theorem for Fields,

$$|G| = |Gal(K/\mathbf{Q})| = (K : \mathbf{Q}) \text{ divides } (K_n : \mathbf{Q}).$$

Again by the Lagrange Theorem for Fields,

$$(K_n : \mathbf{Q}) = (K_n : K_{n-1})(K_{n-1} : \mathbf{Q})$$

and so by induction it suffices to show that G_n is a 2-group, where G_n is the Galois group of K_n/K_{n-1}.

Now $F_n = \mathbf{Q}(\gamma)$ for some primitive element γ. Since $(F_n : F_{n-1}) = 2$, γ is a root of some quadratic polynomial $q(x)$ with coefficients in F_{n-1}. Furthermore if $f(x)$ is the minimum polynomial of γ over $\mathbf{Q}$, then K_n is the splitting field for $f(x)$ and so K_n is generated over $\mathbf{Q}$ by the roots of $f(x)$.

Let β be any root of $f(x)$. Then $\beta = \sigma(\gamma)$ for some $\sigma \in Gal(K_n/\mathbf{Q})$. Thus β is a root of the polynomial $\sigma(q(x))$ with coefficients in $\sigma(F_{n-1})$. Since $F_{n-1} \subseteq K_{n-1}$ and K_{n-1} is a Galois subfield of K_n,

$$\sigma(F_{n-1}) \subseteq \sigma(K_{n-1}) = K_{n-1}.$$

Thus β is a root of a quadratic polynomial $r(x)$ with coefficients in K_{n-1}.

Now if τ is any Galois automorphism in G_n, then τ fixes K_{n-1} and so τ permutes the two roots of $r(x)$. In particular

$$\tau^2(\beta) = \beta \text{ for all roots } \beta \text{ of } f(x).$$

Since K_n is generated over K_{n-1} by the roots of $f(x)$, we see that $\tau^2 = I$ for all $\tau \in G_n$. Then by 22.4, G_n is a 2-group, as desired. ∎

Part 2. *If the Galois group of the minimum polynomial of α over $\mathbf{Q}$ is a 2-group, then α is a constructible number over $\mathbf{Q}$.*

Proof. Let K be the splitting field for the minimum polynomial of α over $\mathbf{Q}$. In order to prove this result, it will suffice to produce a tower of fields

$$\mathbf{Q} = K_0 \subset K_1 \subset \cdots \subset K_n = K$$

with $(K_{i+1} : K_i) = 2$ for all i. In turn, by the Galois Correspondence, it suffices to produce a tower of subgroups

$$G = Gal(K/\mathbf{Q}) = G_0 \supset G_1 \supset \cdots \supset G_n = \{I\}$$

with $(G_i : G_{i+1}) = 2$ for all i.

Indeed if we can produce a normal subgroup H of G of cardinality 2, then K^H is a Galois subfield of K with $(K : K^H) = 2$ and

$$|Gal(K^H/\mathbf{Q})| = |G/H| = \frac{1}{2}|G|.$$

Then we will be done by induction. ∎

Thus we may complete our proof via the following sequence of exercises.

Exercises

22.5. Prove: If G is a finite group with $|G| = 2^n$ for some $n \geq 1$, then G contains an element of order 2. (*Hint*: Pair elements with their inverses, or quote Cauchy's Theorem.)

Definition. Let G be a group. The **center** of G, $Z(G)$, is $\{z \in G : zg = gz \ \forall \ g \in G\}$.

22.6. Prove: $Z(G)$ is a subgroup of G and every subgroup of $Z(G)$ is a normal subgroup of G.

22.7. Prove: If G is a finite group with $|G| = 2^n$ and if $Z(G) \neq 1$, then G has a normal subgroup of cardinality 2.

22.8. Let G be a finite group. For each element $g \in G$ define the conjugation function:

$$c_g : G \to G \text{ via } c_g(x) = gxg^{-1}.$$

Prove: For each $g \in G$, c_g is a permutation of the set G. Moreover c_g is the identity permutation if and only if $g \in Z(G)$.

Thus if G is a finite group of cardinality m with $Z(G) = 1$, then G acts as a group of permutations (subgroup of S_m) on the set G by the conjugation action.

22.9. Suppose that G is a finite group of cardinality 2^n. Prove: $Z(G) \neq 1$ and G has a normal subgroup of cardinality 2. (*Hint*: Suppose that $Z(G) = 1$. The set G is a disjoint union of G-orbits under the conjugation action of the group G on the set G. What does Lagrange's Theorem say about the cardinality of each G-orbit. How many G-orbits have cardinality 1?) ∎

Thus 22.9 completes the proof of the Fundamental Theorem on Constructible Numbers. Notice that this theorem has in particular the following corollary, which does not seem to be at all obvious.

Corollary. *Let α be a constructible number over $\mathbf{Q}$. Let $p(x)$ be the minimum polynomial of α over $\mathbf{Q}$. Then every root of $p(x)$ is constructible over $\mathbf{Q}$.*

We now discuss a way to show that there exist nonconstructible numbers whose minimum polynomial over $\mathbf{Q}$ has degree a power of 2, specifically 4. Namely if

$$p(x) = x^4 + cx^2 + dx + e = (x - r_1)(x - r_2)(x - r_3)(x - r_4)$$

with $c, d,$ and $e \in \mathbf{Q}$, then the numbers

$$t_1 = (r_1 + r_2)(r_3 + r_4), t_2 = (r_1 + r_3)(r_2 + r_4), t_3 = (r_1 + r_4)(r_2 + r_3)$$

CONSTRUCTIBLE NUMBERS AND SOLVABLE EQUATIONS   221

are roots of the Lagrange resolvent cubic equation

$$R(x) = x^3 - 2cx^2 + (c^2 - 4e)x + d^2.$$

Exercises

22.10. For $p(x) = x^4 + x + 1$, find the resolvent cubic $R(x)$.

22.11. For $R(x)$ as in 22.10, prove that $R(x)$ is irreducible over **Q**.

22.12. For $p(x)$ as in 22.10, conclude that if E is the splitting field for $p(x)$ over **Q**, then $(E : \mathbf{Q}) = 12$ or 24. In particular conclude that the roots of $p(x)$ are not constructible numbers. ∎

Galois' work also clarified the question of when a polynomial equation can be solved by a process involving only addition, subtraction, multiplication, division, and extraction of roots. We shall not prove this but only state Galois' fundamental definitions and results.

Definition. A group G is **solvable** if there is a tower of normal subgroups of G:

$$G = G_0 \supset G_1 \supset G_2 \supset \cdots \supset G_n = I$$

such that the quotient groups G_i/G_{i+1} are all abelian groups.

Theorem. *Let $p(x)$ be a polynomial equation with coefficients in **Q**. Then $p(x)$ can be solved by a process involving only addition, subtraction, multiplication, division, and extraction of roots if and only if the Galois group of $p(x)$ is a solvable group.*

Exercises

22.13. Verify that the symmetric group S_n is a solvable group for $n \leq 4$. ∎

On the other hand S_n is *not* a solvable group for any $n \geq 5$. Now it is possible to show that "most" polynomial equations of degree n have S_n as their Galois group. So most polynomial equations of degree n are generic (i.e., have algebraically independent roots), just as Lagrange suspected, and in particular, most polynomials are not solvable by radicals.

22.14. Prove that S_5 is not a solvable group. (*Hint*: Any normal subgroup must be a union of S_5-orbits under the conjugation action defined in 22.8. Moreover $\{1\}$ must be one of these orbits. Now apply Lagrange's Theorem to find all the normal subgroups of S_5.) ∎

As noted earlier Galois' work to a large extent closed the book on the subject of finding algebraic algorithms for solving polynomial equations. Much more important, Galois' work opened the book of group theory and more generally (in conjunction with other work of Gauss) opened a vast and fascinating book of abstract mathematical structures and "Galois correspondences" between them. His work appeared almost impenetrable (or wrong) to his contemporaries, who weren't prepared for it. (Gauss might have grasped it, but he seems never to have looked at it.)

Finally it was clarified by Camille Jordan in his book *Traité des substitutions et des équations algébriques* around 1870. Two young men who came to Paris in the late 1860s—Felix Klein and Sophus Lie—learned Galois' theory from Jordan and were profoundly influenced by it. Klein was led to articulate his Erlanger Programm describing all geometries in terms of the action of a group of isometries on a space. Lie was motivated to search for a Galois Correspondence for Differential Equations, which led him to the important concepts of a Lie group and a Lie algebra.

And so mathematics evolves.

INDEX

A

Abel, Nils Henrik, 56, 188, 217
Angle, 23
Archimedean Property of Real Numbers, 38, 103
Archimedes, 16
Ars Magna, 50
Artin, Emil, 2

B

Binomial Theorem, 123
Bracket product, 100

C

C, 7
Cancellation Property, 53
Cardano's Formula, 51
Cardinality, 9
Cartan, Elle, 100
Cartan-Dieudonne' Theorem, 31
Cartesian product, 7
casus irreducibilis, 54
Cauchy, Augustin Louis, 78
   Cauchy Counting Formula, 86
   Cauchy cycle notation, 80
   Cauchy's Theorem, 140
Cayley, Arthur, 139–140
Characteristic polynomial, 29
Chinese Remainder Theorem, 134–135
Classification of Semisimple Lie Algebras over C, 100
Closed, *See* operation
Complex number, 60
   conjugate of, 60
   norm of, 60
   polar form of, 61
   trace of, 60

Congruence
   of geometric figures, 15, 17
   of numbers, 124
Constructible
   angle, 173
   number, 170
   polygon, 173
Constructions, *See* Straight-Edge and Compass Constructions
Coset (left, right), 137

D

D_n, 33
Da Coi, Zuanne, 57
Da Vinci, Leonardo, 41
Dedekind, Richard, 163
Del Ferro, Scipione, 50
De Moivre, Abraham, 61
   De Moivre's Formula, 61, 63
Descartes, Rene', 7, 123
   Descartes' Factor Theorem, 52–53
   Descartes' Formula, 41–42
Diophantus' Chord and Tangent Method, 56
Disquisitiones Arithmeticae, 170
Division Algorithm
   for numbers, 102
   for polynomials, 117
Divisor, 104
   common, 107
   greatest common, 107
Domain, 7, 114
   of numbers, 110
   unique factorization (UFD), 163
Dual space, 214
   dual basis, 214

E

Eigenvalue, 29
Eigenvector, 29

Eisenstein, F.G., 186
　Eisenstein's Irreducibility Criterion, 186
Elements, The, 46
Elliptic curve, 56–57
　modularity of, 168
Equations
　cubic, 49–56
　quadratic, 47
　quartic, 57–58
Equivalence class, 10
Euclidean Algorithm, 104–106, 142
Euclid's Lemma
　Version 1 (for Numbers), 107
　Version 1 (for Polynomials), 119
　Version 1 (for $\mathbf{Z}[i]$), 144
　Version 2 (for Numbers), 109
　Version 2 (for Polynomials), 119
　Version 2 (for $\mathbf{Z}[i]$), 144
Euler, Leonhard, 122, 124–128, 165
　Euler's Theorem, 127
　Euler–Fermat Theorem, 127
　Euler's function, 126

F

$F(\alpha)$, 181
$F[x]$, 117
$\varphi(n)$, 126
Factor Theorem, 52, 120
Fermat, Pierre de
　Fermat's Last Theorem, 157–168
　　for Polynomials, 158
　　for $n = 3$, 165–168
　　for $n = 4$, 163–164
　Fermat's Little Theorem, 122
　Fermat's Two Squares Theorem, 145
　Method of Infinite Descent, 158–160, 166–168
Ferrari, Lodovico, 57–58
　Ferrari's Method, 57–58
Field, 116
　automorphism of, 187
　extension field, 178
　　degree of, 191
　F-automorphism of, 208
　F-isomorphism of, 208
　finite, 201–203
　Galois field, 209
　isomorphism of, 187
　of numbers, 52
　skew field, 98, 151
　splitting field, 180

Fixed point, 8
Fontana, Niccolo (Tartaglia), 50
Four Squares Theorem, 149, 155–156
Frobenius, F.G., 86
Function, 7–8
　bijective, 9
　composition of, 9
　domain of, 7
　fixed point of, 8
　identity, 9
　injective, 9
　inverse of, 9
　one-to-one, 9
　onto, 9
　range of, 7
　surjective, 9
　target set of, 7
Fundamental Theorem of Algebra, 71–72, 200
Fundamental Theorem of Arithmetic, 110
Fundamental Theorem of Galois Theory, 216
Fundamental Theorem on Constructible Numbers, 218

G

G_a, 84
$\mathrm{Gal}(E/F)$, 208
Galois, Evariste, 1–2, 3, 68, 71
　Galois Correspondence, 213
　　Theorem, 216
　Primitive Element Theorem of, 214–215
Gaussian integers, 65, 142
　associates, 144
　irreducible, 144
　Gaussian prime, 144
　unit in, 144
Gauss, Karl F., 17, 56, 68
　Gauss' Lemma, 185
　Gauss map, 174–175
　Gauss sum, 174–175
gcd, 107, 118
Gibbs, J. Willard, 96
Glide reflection, 30
Golden Rectangle, 105
Golden Ratio, 105
Group, 20
　abelian, 21, 217
　abstract group, 20
　alternating, 77
　automorphism of, 176
　binary icosahedral, 99
　binary octahedral, 99

INDEX   **225**

binary tetrahedral, 99
Cayley table of, 139–140
center of, 220
circle group, 62
cyclic, 36
dihedral, 33
element of, 20
   order of, 36, 127–128
finite, 136
Galois group, 208
homomorphism of, 62
   kernel of, 82
icosahedral, 91
isometry group, 20
isomorphic, 34
isomorphism of, 34
nonabelian, 21
octahedral, 91
of functions, 19
orthogonal, 27
quaternion, 98
quotient group, 218
solvable, 221
subgroup, 21
   index of, 137
   normal, 137, 217
symmetric group, 33
symmetry group of a figure, 32
tetrahedral, 91

H

Hamilton, William Rowan, 53, 95–100, 184
Heath, Sir Thomas, 15
Heptagon, regular, 172–173
Hexagon, regular, 78
Hypercomplex numbers, 98

I

Induction, Principle of Mathematical, 11–12
Isometry, 15
Isomorphism, 9
   isomorphic, 11

J

Jacobi, C.G.J., 56, 100
   Jacobi Identity, 56, 100
Jordan, Camille, 222

K

Khayyam, Omar, 49
Killing, Wilhelm, 100
Klein, Felix, 14, 43, 222
   Klein 4-group, 39–40
Kronecker, Leopold, 71, 185, 209

L

La Ge'ometrie, 53
Lagrange, Joseph Louis, 1, 68
   Lagrange's Theorem
     Orbit-Stabilizer Theorem, 84
     Version 1, 82
     Version 2, 137
     for Field Extensions, 192
Lambert, Johann Heinrich, 195
Leibniz, Gottfried, 16, 61, 64, 68
Lie, Sophus, 100, 222
   Lie algebra, 100
Lindemann, Ferdinand, 195
Linear transformation, 25, 184

M

$M_2(\mathbf{R})$, 66
$M_2(\mathbf{C})$, 97
Map, mapping, 8
Mathematical Induction, Principle of, 11
   Complete Mathematical Induction, 11
Mersenne, Marin, 123
Mobius, A.F., 14, 43

N

$\binom{n}{k}$, 123
Newton, Isaac, 56, 57, 69, 123
Noether, Emmy, 2
Nonagon, regular (9-gon), 173–176
Numbers
   constructible, 170
   greatest common divisor of (gcd), 107
   divisor of, 104
   factor of, 104
   least common multiple of, 111
   multiple of, 104
   one-step constructible, 193
   prime, 108
   relatively prime, 126
   square-free, 148
   transcendental, 195

O

O(2), 27, 32
Operation
    associative, 10
    closed under, 10
    commutative, 10
Orbit ($\sigma-$), 78
    H-orbit, 81
Orthogonal matrix, 27, 88

P

Partition, 10
Pascal, Blaise, 123
Permutation, 8, 33
    cycle decomposition of, 80
    cycle notation for, 80
    cycle structure of, 81
    fixed point of, 8
Petersen's graph, 94
Petry, Jason, 183, 190
Pigeon-Hole Principle, 9
Plato, 16
Platonic solid, 41
    cube, 92–93
    dodecahedron, 92
    icosahedron, 91
    octahedron, 91
    tetrahedron, 91
Pole, 89
Polyhedron, regular, 41
Polynomial
    alternating, 76
    content of, 184
    cyclotomic, 120, 132
    discriminant of, 76
    divisor of, 118
    elementary symmetric, 69
    factor of, 118
    Galois group of, 208
    greatest common divisor (gcd) of, 118
    Irreducible, 119
    minimum polynomial, 182
    monic, 119
    primitive root of, 133
    reducible, 119
    symmetric, 68
    value of, 77
Prime Number Theorem, 108
Primitive Element Theorem, 214–215
Primitive nth root of 1, 131
Pythagorean Theorem, 46
Pythagorean triple, 47–48
    primitive, 111

Q

$\mathbf{Q}$, 7
$\mathbf{Q}(i)$, 117
Quaternions (Hamilton's), 95
    scalar, 96
    vector, 96

R

$\mathbf{R}$, $\mathbf{R}^1$, $\mathbf{R}^2$, 7
Range, 7
Real projective plane, 93
Reflection, 24–25
Relation, 10
    equivalence, 10
    reflexive, 10
    symmetric, 10
    transitive, 10
Ribet, Ken, 168
Rlemann, G.F.B., 14
Ring, 66
    center of, 149
    centralizer of an element of, 152
    characteristic of, 149
    commutative
        ideal of, 197
        principal ideal of, 197
        quotient ring of, 197
    commuting ring of an element of, 152
    division, 98, 151
    homomorphism of, 66
    ideal (2-sided) of, 150
    idempotent in, 147
    left ideal of, 150
    nilpotent element of, 147
    principal left ideal of, 150
    of numbers, 52
    unit group of, 130
Rotation, 25

S

S_n, 38
SO(2), 36
SO(3), 89
Side-Angle-Side Proposition (SAS), 15, 30–31
Spin covering, 99
Stabilizer, 84

Straight-Edge and Compass Constructions, 170
   elementary, 172
Sylow, Ludwig, 141, 203–204
   Sylow's Theorems, 141, 203–204

T

Tartaglia, *See* Fontana, Niccolo
Thales, 15
Tignol, Jean-Pierre, 3
Traite' des Substitutions et des e'quations alge'briques, 222
Translation map, 18, 23
Two Squares Theorem, 145

U

U_n, 128
Unique Factorization Theorem
   for polynomials, 119
   See also Fundamental Theorem of Arithmetic

V

Van der Waerden, B.L., 2

Vector space, 178
   basis of, 179
   dimension of, 180
   isomorphism of, 180
   linearly independent subset of, 179
   spanning set of, 179
   subspace of, 178
Viete, Francois, 54–55

W

Waring, E., 69
Wedderburn, J.H.M., 152, 206
   Wedderburn's Theorem, 152, 206
Well-Ordering Principle, 36, 103
Weyl, Hermann, 2, 41
Wiles, Andrew, 56, 122, 158, 168

Z

Z, 7
Z[i], 65, 142
   Division Algorithm in, 142
Z[i]/n**Z**[i], 146
Z/n**Z**, 129–130
Zsigmondy's Theorem, 153, 205–206